STEEL STRUCTURES

STRUCTURES

PRACTICAL DESIGN STUDIES

STEEL STRUCTURES

PRACTICAL DESIGN STUDIES

T. J. MacGinley

Principal Lecturer
Sunderland Polytechnic

LONDON AND NEW YORK
E. & F. N. SPON

First published 1981 by
E. & F. N. Spon Ltd
11 New Fetter Lane, London EC4P 4EE
Published in the USA by
E. & F. N. Spon
in association with Methuen, Inc.
733 Third Avenue, New York NY 10017

Printed in Great Britain by
Richard Clay (The Chaucer Press) Ltd., Bungay, Suffolk

ISBN 0 419 12560 4 (cased)
ISBN 0 419 11710 5 (paperback)

British Library Cataloguing in Publication Data

MacGinley, T. J.
   Steel structures.
   1. Steel, Structural
   2. Structural design
   I. Title
   624.1'821   TA684

   ISBN 0-419-12560-4 (cased)
   ISBN 0-419-11710-5 (paperback)

Contents

Preface

The purpose of the book is to present the principles and practice of design for some of the main modern structures. It is intended for final year degree students to show the application of structural engineering theory and so assist them to gain an appreciation of the problems involved in the design process in the limited time available in college. In such a presentation many topics cannot be covered in any great detail.

Design is a decision-making process where engineering judgement based on experience, theoretical knowledge, comparative design studies, etc, is used to arrive at the best solution for a given situation. The material in the book covers the following:

(a) discussion of conceptual design and planning;
(b) presentation of the principles and procedures for the various methods of analysis and design;
(c) detailed analysis and design for selected structures. Preliminary design studies are made in other cases where the full treatment of the problem is beyond the scope of this book.

In detailed design, the results are presented in the form of sketches showing framing plans, member sizes and constructional details.

Although the book is primarily concerned with the design of steel structures, important factors affecting both the overall design and detail required are discussed briefly. These include the choice of materials, type of foundations used, methods of jointing, the fabrication process and erection methods. Other design considerations such as fatigue, brittle fracture, fire resistance and corrosion protection are also noted.

The use of computers in design is now of increasing importance. Where required, computer programs are used in the book for analysis. While examples of computer-aided design have not been included, a project on this topic is listed at the end of the book. It is felt that the student must thoroughly understand design principles before using design programs.

In college, the student is instructed through formal lectures backed by reading from textbooks and journals and by consultation with staff and fellow students. The acquisition of knowledge and the exchange of ideas help him to develop his expertise and judgement and to make sound decisions. However, the most important part of the learning process is the carrying out of practical design work where the students are given selected coursework exercises which cover the stages in the design process. Such exercises have been included at the end of most chapters. These, generally, consist of making designs for given structures including framing plans, computer analysis, design and detail drawings.

In many first degree courses, the student is also required to undertake a project for which he may choose a topic from the structural engineering field. This gives him the opportunity to make a study in a particular area of interest in greater depth than would be possible through the normal lectures. Some suggestions for projects are given at the end of the book. These may be classified as follows:

(a) comparative design studies;
(b) computer-aided design projects;
(c) construction and testing of structural models and presentation of results in report form.

The intention of the book is to help equip the young engineer for his role in structural engineering in industry. It is important to foster interest in structural design where this is shown by a student. It is hoped that this book will go some way towards this goal.

Acknowledgements

The author gratefully acknowledges the following.

The British Standards Institution: Extracts from British Standards are reproduced by permission of the British Standards Institution, 2 Park Street, London W1A 2BS, from whom copies can be obtained.

Her Majesty's Stationery Office: Material is included from the following two publications, *The Building Regulations* (1976), and Newberry, C. W. and Eaton, K. J. (1974) *Wind Loading Handbook*, Building Research Establishment. The material from these publications is Crown copyright and is reproduced by permission of the Controller, H.M.S.O.

Building Research Establishment: Material is included from the following publication with permission. Wood, R. H. (September, 1974) *Effective lengths of columns in multi-storey buildings*, Current paper CP85/74. This was also published in the *Structural Engineer* in July, August and September, 1974.

Constructional Steel Research and Development Organisation (Constrado): By permission, Chapter 7 of this book about plastic design is based on Morris, L. J. and Randall, A. L. (1975) *Plastic Design*, (1979) *Plastic Design – supplement*.

British Steel Corporation, Tubes Division: Material is included from the following publication with permission. *Nodus Space Frame Grids* (1974), *Part 1 Design, Part 2 Analysis, Part 3 Construction*.

Finally, I must acknowledge a great debt to my wife, Trudy, who typed the manuscript and assisted in checking the proofs. Without her help, this book could not have been written.

Notation

A	critical stress — bending stresses for plate girders
A	greater projection of a base plate beyond the stanchion
a_1	net sectional area of the connected leg of an angle
a_2	sectional area of the unconnected leg of an angle
B	lesser projection of a base plate beyond the stanchion
C_s	critical stress — bending stresses for plate girders
D	overall depth of a beam or plate girder
d, d_1	clear depth of web
d_3	clear depth of web of a universal beam between root fillets
f_{bc}	compressive stress due to bending
f_{bt}	tensile stress due to bending
f_c	axial compressive stress
f_q'	average shear stress calculated on the web of a beam
l	effective length of the compression flange of a beam or the effective length of a column
L	length of a column from centre-to-centre of intersections with supporting members, length of beam or cantilever (Subscripts $1, 2 \ldots$)
p_b	allowable bearing stress
p_{bc}	allowable compressive stress in bending
p_{bt}	allowable tensile stress in bending
p_{bct}	allowable bending stress in base plates
p_c	allowable compressive stress in an axially loaded column
p_q'	allowable average shear stress in the web of a beam
p_t	allowable axial tensile stress
r	radius of gyration
r_Y	radius of gyration of a beam section about its axes lying in the plane of bending
T	mean thickness of flange
t	thickness of web or baseplate
w	pressure on the underside of the base
W	total load on a beam or purlin

Standard symbols from CP3: Chapter V: Part 2
Wind loads

A	area of surface
A_e	effective frontal area of the structure
b	breadth of building normal to the wind
C_f	force coefficient
$C_{f'}$	frictional drag coefficient
C_{pe}	external pressure coefficient
C_{pi}	internal pressure coefficient
d	depth of building in direction of wind
F'	frictional drag
H	height above ground of building or part considered
h	height to eaves or parapet
l	length or greater horizontal dimension of building
q	dynamic pressure
S_1	topography factor
S_2	factor for ground roughness, building size and height above ground
S_3	statistical factor giving degree of security
V	basic wind speed
V_s	design wind speed
w	width or lesser horizontal dimension of building
α	angle of wind to the axis of the building

Standard symbols from the draft standard specification for the structural use of steelwork in building

M_c	moment capacity
p_y	design strength
S	plastic modulus of a section
u_s	specified minimum ultimate tensile strength
y_s	specified minimum yield stress of steel
Z	elastic modulus of a section
γ_f	partial factor of safety for loads
γ_l	partial factor of safety for variability of loads
γ_m	partial factor of safety for material strength
γ_p	partial factor of safety for structural performance
λ	slenderness ratio

Other symbols used

A	area of section
A_{net}	tensile area of a bolt at the root of the thread

a	dimension
b	dimension, breadth
D	overall depth of a beam, depth of space grid
E	Young's modulus
F	loads, forces in chords of trusses or space grids. Subscripts $(1, 2 \dots)$
f	actual or calculated stresses. Subscripts $(c, bc, \dots)$
H	horizontal loads or reactions. Subscripts $(1, 2, A \dots)$
M	moments. Subscripts $(1, 2, A, \dots)$
M_p	full plastic moment due to the factored loads
m_p	plastic moment due to the unfactored loads
n	ratio of stresses for reduced plastic modulus of section
	top chord module dimension for a space grid
P, p	loads, forces in chords of trusses. Subscripts $(1, AB, \dots)$
p	allowable stresses. Subscripts $(c, bc, \dots)$
p_{cb}	allowable compressive stress in concrete in bending
p_{max}	maximum bearing pressure
θ, ϕ	angles.

Glossary

1 Structural Elements, Joints and Bases

Bases
Transfer loads from columns to the foundations.
  Bases may be:
  pinned — transmitting thrust and shear;
  fixed — transmitting thrust, shear and moment.
  The main types are:
  slab — consists of a thick plate welded to the column;
  gusseted — consists of stiffened arrangement of plates;
  pocket — column section is embedded in the concrete.

Beams and plate girders
Horizontal members resisting loads in shear and bending. Types are:
  beams — usually rolled universal beam sections;
  compound beams — universal beams strengthened by the addition of flange plates. They are used where the depth is restricted;
  plate girders — these are large beam members fabricated from plates by welding. The webs require stiffening. They are used to carry heavy loading over long spans.

Bracing
Single or double diagonal members which form trusses with columns or beams to provide stability and resist horizontal load. Angles, channels and structural hollow sections are used.

Columns, stanchions or struts
These are members, generally vertical, which primarily resist axial load. They are more often subjected to thrust and moment. The usual section used is the rolled universal column. Other types are:
  laced and battened columns — consists of two sections connected together by lacing or batten plates;
  built up and box columns — fabricated from plates.

Crane girders

These resist vertical and horizontal loads from cranes. They usually consist of a universal beam with a channel, flanges down, welded to the top flange. Plate girders are required for heavy cranes where long spans are involved.

Joints

These connect individual members together. Joints may be pinned or rigid. Types are:
  bolted – using black bolts or high-strength friction grip bolts; or
  welded – with fillet or butt welds.

Purlins

Beam members carrying roof sheeting.

Sheeting rails

Beam members carrying side sheeting.

Ties

Tension members. They are generally truss members or hangers in suspended structures.

Trusses and lattice girders

These are framed assemblies generally carrying lateral load. The individual members are in tension or compression. Trusses are fabricated from angles, tees, structural hollow sections etc.

2 Structures

Steel frameworks consisting of beams, columns, trusses, bracing, joints and bases carrying the building loads. Structures are classified firstly into:

Plane frames

where the members lie in one plane such as with portal frames and truss and stanchion frames;

Space frames

where the three dimensional structure is treated as a unit for analysis and design, e.g. space decks and domes.
A further classification is into:
  braced frames – where the joints are generally pinned and bracing is required to provide stability. These frames are usually arranged to be statically determinate;
  rigid frames – where the joints transmit moment without relative rotation between the members meeting at a joint. These structures are statically indeterminate.

3 Analysis

This is the process used to determine the moments, thrusts and shears at appropriate points in the frame for design. Elastic and plastic methods are used. The first classification for elastic methods is:

Statically determinate methods
for trusses and braced frames such as force diagrams, simply supported beam systems etc.

Statically indeterminate methods
for rigid frames. These are divided into:
  (i) *force or flexibility method* where the frame is released and values of the redundants required to restore continuity are determined. Some methods here are strain energy and moment area;
  (ii) *displacement or stiffness method* where the frame is restrained against movement. The restraints are released and values of rotations and deflections are determined and used to find the forces and moments in the frame members. Methods of this type are slope deflection and moment distribution.

Matrix methods
The analysis is formulated using matrix alegebra. Computer analysis programs are based on the matrix stiffness method.

Plastic analysis
The collapse load, the number and location of plastic hinges to cause collapse are determined.

4 Design methods for steel frame structures

Simple design
Joints are taken as pinned. Bracing is required to give stability.

Semi-rigid design
Joints between members transmit some moment which is taken into account in the analysis.

Rigid design
No relative rotation between members at joints occurs.
  Elastic design – uses elastic analysis and member design. Allowable stresses must not be exceeded.
  Plastic design – uses plastic analysis and member design. The design is such as to give an adequate load factor against failure.

Limit state design

The new steel code will be to limit state theory. This method is based on the actual behaviour of materials and structures using partial factors of safety for materials and loads. All limit states or the reasons for the structures becoming unfit for use must be investigated.

5 Design process

Actual stress

The load per unit area. This may be the calculated axial, bending, shear, torsion or bearing stress. The separate stresses are combined by various methods.

Allowable stress

The yield stress or critical buckling stress divided by a factor of safety. This stress must not be exceeded at working loads. Allowable stresses are given in BS449.

Buckling (instability)

Movement away from the initial or unloaded position. This is the cause of failure in slender or unrestrained members at low loads.

Crushing

Occurs when the material flows plastically as the yield stress is reached.

Effective length

The actual or theoretical length of a column between points of contraflexure at the instant of buckling. Alternatively, it is the length of a pin-ended column which can carry the same critical load as the actual column. The term is also used in respect of the buckling length of an unrestrained compression flange of a beam.

Factored load

The working load multiplied by the load factor. It is the ultimate load applied to the frame and causes the formation of sufficient hinges for collapse to occur.

Interaction formula

This combines the effects of different actions in order to predict the performance of a member, e.g. combining effects of axial load and bending moment in design of a beam-column.

Lateral restraints

Members or stays from other members that restrain a column or compression flange of a beam from buckling.

Load factor
The collapse load divided by the working load. It gives the margin of safety for the frame.

Slenderness ratio
The effective length divided by the radius of gyration. It is a measure of the likelihood of a member to buckle.

Plastic hinge
Points of maximum or plastic moment where the entire section is plastic and large rotation can occur without increase in moment.

Plastic moment
The moment developed at the plastic hinge when all stresses in the section are at the yield stress.

1 Structural engineering, design principles and methods

1.1 Structural engineering and the design process

In general, structural engineering covers planning, design and construction of all structures. For steelwork these include self-supporting and load-bearing forms consisting of frameworks, plated structures, shells and tension structures. In particular the aim of structural design is to produce the design and drawings for a safe and economical structure that fulfills its required purpose. The steps in the design process can be set out as follows.

(a) Conceptual design and planning. This involves selecting the most economical structural form and materials to be used. Preliminary designs are often necessary to enable comparisons to be made.
(b) Detail design for a given type and arrangement of structure. This includes:

   (i) idealization of the structure for analysis and design;
   (ii) estimation of loading;
   (iii) analysis for the various load cases and combinations of loads and identification of the most severe design actions;
   (iv) design of the foundations, structural frames, elements and connections;
   (v) preparation of the final arrangement and detail drawings.

The materials list, bill of quantities and specification may then be prepared to enable the estimates and tender documents to be completed.

The structural designer uses his knowledge of structural mechanics and design, materials, geotechnics, and the codes of practice and combines this with his practical experience to produce a satisfactory design. He takes advice from specialists and makes use of design aids, handbooks and computers to help him in making decisions and to carry out complex analysis.

1.2 The qualities and role of the structural engineer

The above description infers that the designer's work is mechanical in nature. It is useful to consider his position in building construction where the parties involved are:

client who consults an architect about his requirements;

architect who plans and controls the project and engages the consultants;
consultants who carry out the design, prepare drawings and tender documents
    and supervise construction
contractor who carries out fabrication, construction and installation of equip-
    ment.

The designer is also a member of a team at design office level consisting of: chief
engineer, project engineer, design engineer, computer staff, technician engineers
and specialists. He must fit smoothly into the team.

Some of the qualities and attributes needed by the structural engineer to
operate successfully are: flair, sound knowledge and judgement, experience and
exercise of great care. His role may be summarized as planning, design, preparation of drawings and tender documents and supervision of construction. He
makes decisions about materials, structural form and design methods to be used.
He recommends acceptance of tenders, inspects, supervises and approves fabrication and construction. He has an overall responsibility for safety and must
ensure that the consequences of failure due to accidental causes are limited in
extent.

The designer's work which is covered, in part, in this book is one portion of a
structural engineer's work.

1.3 Classification of structures

Structures are needed for the following purposes:

(a) to enclose space for environmental control;
(b) to support people, equipment, materials etc at required locations in space;
(c) to contain and retain materials;
(d) to span gaps for the transport of people, equipment, materials etc.

Framed buildings may be classified according to use:

domestic – houses and flats;
commercial – offices, banks, department stores, shopping centres etc;
institutional – schools, hospitals, gaols etc;
exhibition – churches, theatres, museums, leisure centres, sports buildings etc;
industrial – factories, warehouses, power stations, steelworks, manufacturing
    plant, aircraft hangers etc.

Other important engineering structures are:

bridges – truss, girder, arch and suspension;
towers – water towers, pylons, lighting towers etc;
offshore structures – oil well platforms;
special structures – grandstands, multi-storey carparks, radio telescopes, mine
    headframes etc.

Each type of structure in the above list can be constructed using a variety of materials, structural forms or systems. The engineer often uses a classification for steel structures based on the form or system used. This gives:

(a) single storey single- or multi-bay structures which may be either truss and stanchion frames or rigid frames;
(b) multi-storey single- or multi-bay structures which may be either braced or rigid frame construction;
(c) space structures – space decks, domes, towers, etc. Space decks and modern domes are redundant structures. Towers may be statically determinate space structures;
(d) tension structures and suspended structures;
(e) stressed skin structures.

Illustrations of some of the steel structures listed above are shown in Fig. 1.1. Fig. 1.1(a) and (b) show, for comparison, braced traditional construction and the rigid frame alternative for single-storey industrial buildings and multi-storey buildings, respectively. The sawtooth roof structure and space deck for covering large areas are shown in Fig. 1.1(c). Only framed structures are dealt with. Plate and shell type structures, e.g. steel tanks, are not considered here. Detail designs and design studies are given in the book for selected structures.

1.4 Conceptual design, innovation and planning

Conceptual design is the function of choosing a suitable form or system or arrangement to meet a given structural situation. This function is often the sole prerogative of the architect. Ideally, conceptual design should result from a team effort where architect, civil, structural and service engineers all contribute to the final solution. Modern consulting practices take this multi-disciplinary approach to conceptual design. However, the greatest achievements often are made by gifted individuals. For example, to name two individuals:

Fazlur Khan – tube system for tall buildings;
Buckminster Fuller – geodesic domes.

In the United Kingdom it is often the architect who produces the form and arrangement he considers as the best solution to a given problem. He bases his decision on functional, aesthetic, environmental and economic considerations. The structural engineer is then faced with choosing and designing a structural system that will bring the architect's ideas into being and he may be hampered by not having had a hand in the original decisions. Any of the factors mentioned may be of paramount importance in arriving at a decision, for example, in industrial plant, the functional requirement controls, whereas in an exhibition building the aesthetic aspect is a major consideration.

Novelty and innovation are always desirable and we seem to strive after these

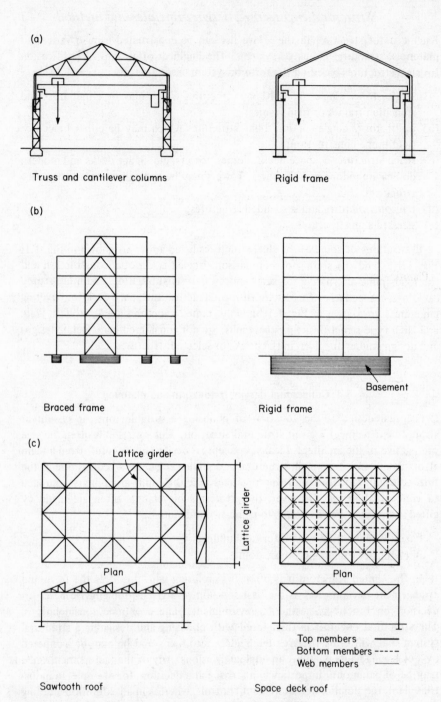

(a)

Truss and cantilever columns   Rigid frame

(b)

Braced frame   Rigid frame

Basement

(c)

Lattice girder

Lattice girder

Plan   Plan

Top members ———
Bottom members -----
Web members ———

Sawtooth roof   Space deck roof

Figure 1.1 (a) Single-storey single bay industrial buildings. (b) Multi-storey buildings. (c) Roof structures covering large areas.

goals. It should be realized that man has always exploited the forms of construction possible within the limits of the materials available and the state of his knowledge at the time. In the past as well as in the present, structural failures determine when limits are reached. Often it is not a new solution that is required but the correct choice and use of a well proved existing structural system that gives the best answer. The engineer seeks new and improved methods in analysis, design and construction through his research and development. These advances lead to safer and more economic structures. The following are instances of recent structural engineering innovation:

(a) analysis – matrix and finite element methods;
(b) design – plastic design, limit state theory and optimization
(c) construction – space decks, geodesic domes and the modern tension structures, box girder bridges, suspended multi-storey buildings etc.

Planning may be described as the practical expression of conceptual design. The various proposals must be translated into drawings, such as general arrangements consisting of plans and elevations to set out the functional requirements and perspective drawings to give a realistic impression of the finished concept. Scale models of the complete structure in its surroundings may be made to assist in making the final decision. The preparation and presentation of planning proposals are very important as the final approval for schemes often rests with non-technical people.

The engineer must also consider construction in materials other than steel such as structural concrete, masonry or timber and make the most appropriate selection. A list of factors that need to be considered at the conceptual and planning stage would include the following:

(a) the location of the structure and environmental conditions;
(b) the site and foundation conditions;
(c) the weather conditions likely during construction;
(d) the availability of materials;
(e) the transport of materials and fabricated elements to site;
(f) the quality of workmanship required and availability of labour for fabrication and erection;
(g) the degree of supervision needed in fabrication and erection stages;
(h) the likelihood of damage or failure due to fatigue and brittle fracture;
(i) the measures needed to give protection against corrosion and fire;
(j) the possibility of accidental damage;
(k) the maintenance required after completion;
(l) the possibility of demolition in the future.

The final decision on the form and type of structure and construction method depends on many factors and will often be taken on grounds other than cost.

1.5 Comparative designs and optimization

1.5.1 General considerations

Preliminary designs to enable comparisons and appraisals to be made will often be necessary during the planning stage in order to establish which of the possible structural solutions is the most economical. Information from the site survey will be required because foundation design will have a great effect on the type of super-structure to be used as well as the overall cost.

Arrangement drawings, showing the structural system to be used, are made for the various proposals. Then preliminary analysis and designs are carried out to establish the member sizes and weights so that costs of materials, fabrication, construction and erection can be estimated. Maintenance costs must also be taken into account. However, it must be emphasized that it is very difficult to get true comparative costs for different designs. Estimates based on available unit costs can be unreliable when applied to new structural situations. It is generally not possible to obtain costs from contractors at the planning stage. Design comparisons are discussed in more detail below. Worked examples in the book have been selected to show various design comparisons using different structural types or design methods.

By optimization is meant the use of mathematical techniques to obtain the most economical design for a given structure. The aim is usually to determine the topology of the structure, arrangement of floors, spacing of columns or member sizes to give the minimum weight of steel or minimum cost. To apply the method other than in certain simple cases, requires the use of a computer program to carry out the search technique. Optimization is not of great use at present in structural engineering as many important factors cannot be taken satisfactorily into account.

The design of individual elements may be optimized, e.g. plate girders or trusses. However, with optimum designs, the depth of these elements is often some 50 per cent greater than depths normally adopted.

1.5.2 Aims and factors considered in design comparisons

The aim of the design comparison is to enable the designer to ascertain the most economical solution that meets the requirements for the given structure. All factors must be taken into consideration. A misleading result can arise if the comparison is made on a restricted basis. General factors to be considered are:

(a) the materials to be used;
(b) the arrangement and structural system to be adopted;
(c) the fabrication and type of jointing;
(d) the method of erection of the framework to be used;
(e) the type of construction for the floors, walls, cladding and finishes;
(f) the installation of plant, equipment, services etc;
(g) the corrosion protection required;

(h) the fire protection required;
(i) operating and maintenance costs.

Aesthetic considerations are important in many cases and the design adopted is not always based on cost considerations.

Most structures can be designed in a variety of ways. The possible alternatives that may be used include:

(a) the different methods of framing to achieve the same structural solution;
(b) the selection of spacing for frames and columns;
(c) the various alternatives that may be used to stabilize the building and provide resistance to horizontal loading;
(d) the different design methods that may be applied to the same structural form, e.g. simple design or semi-rigid design or rigid design using either elastic or plastic theory;
(e) design in different materials, e.g. mild steel or high-strength steels. Here, the weight saving is offset by the higher cost of the stronger steel.

It should be noted that often no one solution for a given structure ever appears to dominate to the exclusion of all other alternatives. This illustrates the fact that many variables influence decisions.

1.5.3 Specific basis of comparisons for common structures

In the following sections a classification is given on which design comparisons for general purpose buildings and structures may be made. The list is not intended to be exhaustive. Design comparisons for roof structures are discussed in Chapters 8 and 9 and for tall buildings in Chapter 10.

(a) Single-storey single-bay buildings

The plan size span x length is fixed by the client's requirements. The designer can then choose from the following list of variations.

Type of building — and design methods. See Fig. 1.2(a)
 (i) The truss and stanchion frame with cantilever columns, or knee braces with pinned or fixed bases using simple design.
(ii) The three-pinned portal of I section or lattice construction using simple design.
(iii) The rigid portal, pinned or fixed bases in I or box section or in lattice construction. Elastic or plastic design methods may be used. The design may be fully welded or the rigid joints may be made with high-strength friction grip bolts.

Design variables
The basic variable is the column spacing. This affects the size of purlins, sheeting rails, main frame members and foundations. Designs may be made with various

(a)

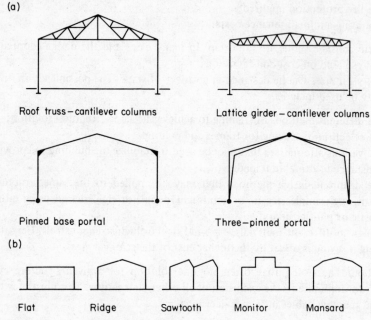

Roof truss – cantilever columns   Lattice girder – cantilever columns

Pinned base portal   Three – pinned portal

(b)

Flat   Ridge   Sawtooth   Monitor   Mansard

Figure 1.2   (a) Different structural frames. (b) Different roof shapes.

column spacings to determine which gives the most economic result. Various roof shapes are also possible. Some types used are flat, ridge, sawtooth, monitor, mansard. See Fig. 1.2(b). The roof slope is a further variable which is usually fixed arbitrarily. Present practice is to use flatter slopes sufficient to ensure run off. Flat roofs or single-slope roofs can also be used. In the longitudinal direction these buildings are in braced simple design. Rigid framing can be used as an alternative to give stability.

(b) Single-storey multi-bay buildings
The comments from (a) apply in this case. The span of the bays is taken to be fixed by the client's requirements.

(c) Multi-storey buildings
The column spacing may be varied in two directions. The location of the liftshaft and stairs may be varied. Not all columns may be continuous throughout the structure. Plate girders can be used to carry upper floor columns over clear areas. Economy in floor steel can be achieved if the bottom storey columns are set in allowing the floor beams to cantilever out.

The type of flooring and arrangement of floor framing adopted affects the overall design of the building. The main types of flooring used are cast *in situ* concrete in one-way or two-way spanning slabs or precast one way floor slabs. The cast *in situ* slabs can be constructed to act compositively with the steel beams.

Various systems or structural combinations may be used to stabilize the building and resist horizontal loads. The building may be braced in both directions, rigid one way and braced the other way, or rigid in both directions. Alternatively, lift shafts, stair walls or special shear walls may be used to provide stability.

For a given framing system various design methods can be used. These are simple design, semi-rigid or rigid design.

Fire protection is necessary with these buildings and solid casing of beams and columns may be taken into account in the structural design. However, light-weight hollow casing is generally used in modern practice.

The type of foundation required affects the choice of form of the super-structure. A common case in point is where provision must be made for differen-tial settlement. Here buildings of simple design perform better than those of rigid design. If a monolithic raft or basement foundation is provided the super-structure can be designed independently of the foundations.

(d) Special purpose structures

In these situations there may be two entirely different ways of framing the struc-ture while complying with all requirements for the finished structure. One such structure is the sports pavillion or grandstand where cantilever construction can be used throughout as one possible solution. Alternatively, end columns sup-porting a lattice girder at the front can be used to carry the front edge of the roof. See Fig. 1.3.

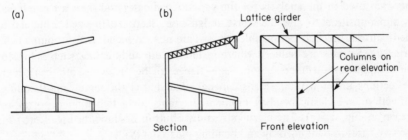

Figure 1.3   (a) Cantilever construction. (b) Lattice girder construction.

1.6 Design theories — elastic, plastic, limit state

In this section, general principles of design in accordance with the various theories are discussed. Particular methods for design of structures set out in British Standards are given in detail in Sections 1.7 and 1.8.

1.6.1 Elastic theory

This is the traditional method of design. The behaviour of steel when loaded below the yield point is very much closer to true elastic behaviour than that of

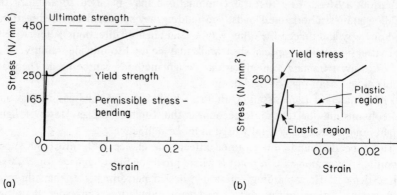

Figure 1.4   (a) Stress–strain curve for Grade 43 steel. (b) Stress–strain curve assumed for plastic design.

other structural materials. Refer to the stress strain diagram for mild steel shown in Fig. 1.4(a). Consequently design based on elastic theory has much to commend it. The structure is assumed to obey Hooke's Law and recover to its original position on removal of the load. Design to elastic theory is carried out in accordance with BS449 *The use of structural steel in building*.

The structure is loaded with the working loads which are taken to mean the maximum loads to which the structure will be subjected during its life. Elastic theory is used in the analysis for the various load cases and these are combined by super-position to give the worst design case. Sections are sized using elastic theory to ensure that permissible stresses are not exceeded at any point in the structure. Stresses are reduced where instability due to buckling, such as in slender compression members, compression flanges of beams, webs, etc. can occur. Deflections under working loads can be calculated at the same time. Loading, deflection and elastic bending moment diagrams for a two pinned portal are shown in Fig. 1.5(a). The elastic stress distribution is shown in Fig. 1.5(b) for the two cases of bending only and bending plus axial load.

The permissible stresses are obtained by dividing the yield stress by a factor of safety. For direct tension the factor of safety is 1.72 and for bending of rolled sections 1.63 for Grade 43 steel. Where stability is a problem, the permissible stress is obtained by dividing the elastic critical buckling stress by a factor of safety, usually 1.7. This one factor takes account of variations in strengths of materials and fabricated members, possible overloads etc. Permissible stresses are given in BS449.

1.6.2 Plastic theory

When a steel specimen is loaded beyond the yield point the strain increases for a period with no further increase in load. This is shown in Fig. 1.4(b) where the

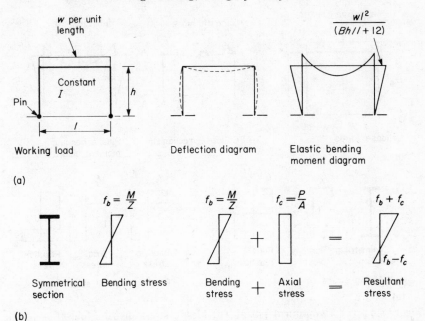

Figure 1.5   (a) Two-pinned portal. (b) Elastic stress distribution. M = binding moment, A = area section, f_c = axial stress, p = axial load, Z = elastic modulus, f_b = bending stress.

strain scale has been increased. The performance of sections in bending at collapse is based on this stress—strain curve. Thus as the moment increases, stresses, which are at first elastic, increase to reach yield at the extreme fibre where the strain continues to increase while internal fibres successively reach yield until a plastic hinge has formed as shown in Fig. 1.6(b).

Plastic analysis is based on determining the least load that will cause the structure to collapse. The safe load is then the collapse load divided by a load factor. For frames, collapse occurs when sufficient hinges have formed to convert the structure to a mechanism, subject to the condition that the plastic moment or yield stress is not exceeded at any point in the frame.

In design, the structure is loaded with the collapse or factored loads obtained by multiplying the working loads by the load factor and analysed plastically. The loading collapse mechanisms and plastic bending moment diagrams for a two-pinned portal are shown in Fig. 1.6(a). Sections are designed using plastic theory and the stress distributions for sections subjected to bending only and bending and axial load are shown in Fig. 1.6(b). Sections require checking to ensure that local buckling does not occur before a hinge can form. Bracing is required to prevent overall buckling of the member. Separate load cases cannot be combined; the frame must be analysed for each load case. Under working loads the structure is in the elastic region and deflection must be found using

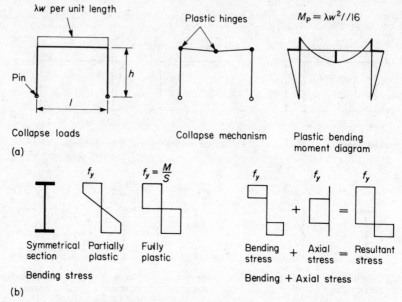

Figure 1.6 (a) Two-pinned portal. (b) Plastic stress distribution. M_p = plastic moment, f_y = yield stress, w = working load, λ = load factor, S = plastic modulus.

elastic theory. Methods are available for estimating the deformed shape of the structure when collapse is imminent.

The method of design is included in BS449 but the load factor to be used is not specified. This must be taken from other literature, such as Constrado publications [7]. A factor of 1.7 is used for combinations of dead and imposed loads and 1.3 where wind load is included.

1.6.3 Limit state theory
Limit state theory originated with the C.E.B. (Comitée European Du Beton) where it was put forward as a more realistic basis for the design of structural concrete than the elastic method then in use. The limit state method has now been widely accepted as a basis of design for all structural materials. The new draft code for structural steelwork is a limit state code [4, 11]. This was issued for comment in 1978 but has since been withdrawn.

The following concepts are central to limit state theory.

(a) Account is taken in design of all the separate conditions that could make the structure unfit for its intended use. These are the various limit states which apply in the given case. These are listed below.
(b) The design is based on the actual behaviour of materials in structures and performance of real structures established by tests and long-term observa-

tions. Good practice must be followed in order that some limit states are not reached.

(c) The overall intention is that design is to be based on statistical methods and probabilistic concepts. It is recognized that no design can be made completely safe but only that a low probability that the structure will not reach a limit state can be achieved. However, full probabilistic design is not possible at present.

(d) Separate partial factors of safety for loads and materials are introduced. This permits a better assessment to be made of uncertainties in loading, variations in material's strengths and the effects of initial imperfections and errors in fabrication. The factors also give a reserve of strength against failure.

The limit states for structural steelwork are:

(a) ultimate limit states
    overall stability
    strength taking account of instability in sections and slender members
    fatigue failure
    brittle fracture
    fire damage leading to failure;
(b) serviceability limit states
    deflection
    vibration
    corrosion
    fire damage leading to loss of serviceability.

In limit state philosophy, the steel code is a Level 1 safety code. This means that safety or reliability is provided on a structural element basis by specifying partial factors of safety for loads and materials. All the relevant separate limit states must be checked. Level 2 is partly based on probabilistic concepts and gives a greater reliability than Level 1 design. Level 3 is a fully probabilistic design for the complete structure.

For materials, the characteristic strength is defined as that value below which not more than 5% of test results fall. Statistical methods are well established in quality control of materials. The partial factor of safety for strength, γ_m, allows for the variation between the strength of material in the structure and the strength specified for design. The draft code defines

Design strength $p_y = \gamma_m$ x characteristic strength

$$= 0.93y_s \text{ but not greater than } 0.73u_s$$

where y_s is the specified minimum yield strength, and u_s the specified minimum ultimate tensile strength. The γ_m factors are incorporated in the design strengths for various steels given in Table 5.7.1 of the draft code. Some values for design strengths are given in Table 1.1

Table 1.1   Design strengths for steel to BS4360

Grade	Thickness (mm)	Design strength p_y (N/mm^2)
43 sections	$\leqslant 40$	240
43 plates	$\leqslant 40$	220
50 sections	$\leqslant 25$	340
50 plates	$\leqslant 63$	320

Ideally characteristic loads would be defined as loads that would have a 95% statistical probability of not being exceeded during the life of the structure. However, statistically derived characteristic loads are not yet available. At present the values specified in CP3, Chapter V, Part 1, for dead and imposed loads are used as the characteristic loads. The wind loads specified in CP3, Chapter V, Part 2, are based on wind speeds which have been derived statistically. The factored load is used in design and this is defined in the draft code:

Factored load = γ_f x specified or characteristic load, where γ_f, the partial factor or safety, is the product of two factors: γ_l allows for the variability in loading from the specified value (this factor depends on the type of load and load combination); γ_p allows for the variability in structural performance. This includes for dimensional variations, quality of workmanship and the difference between actual behaviour and the behaviour of the idealized structure assumed for analysis. The partial factors of safety for loads are given in Table 5.2.1 of the draft code. Part of this table is given here as Table 1.2.

Table 1.2   Partial factors of safety of loads

Type of load or combination		Factor
Dead	maximum	1.4
	minimum	1.0
	minimum for pattern loading	1.2
Imposed load in the absence of wind		1.6
Wind load acting with dead load only		1.4
Wind and imposed loads acting in combination		1.2

The design is made for factored loads using the design strengths of the materials. Sections in bending must satisfy both elastic and plastic criteria. Thus the moment capacity, M_c, for a compact beam section is given by

$$M_c = p_y S \text{ but } \not> 1.25 p_y Z$$

where p_y is the design strength, S the plastic modulus, and Z the elastic modulus.

In design of struts account is taken of geometric imperfections and residual stresses due to the method of manufacture whether it is a rolled section or is fabricated by welding and the thickness of material. This is done by selecting an appropriate value of the Robertson constant from a table. The compressive strength for a material with design strength p_y depends on the Robertson constant and slenderness ratio λ. A more rigorous method is adopted for the design of members subjected to axial load and moment than the interaction formula given in BS449.

In general, the new code will be much more specific and thorough in the design of structural elements.

1.7 Design methods from BS449

Design to BS449, Part 2, 1969, *The use of structural steel in building*, may be made by any of the following methods as set out in Clause 9a.

Simple design – elastic theory
Semi-rigid design
Rigid design – elastic or plastic theory
Experimentally based design.

The clause begins by stating that any part of the structure must be capable of sustaining the most adverse combination of static and dynamic forces which may reasonably occur without the permissible stresses being exceeded. The design methods are now set out briefly below:

1.7.1 Simple design to elastic theory
This is to be used for structures in which the end connections are such that they will not develop restraint moments adversely affecting the members and structure as a whole. The structure is to be assumed pin-jointed for design purposes. The following assumptions are made:

(a) beams are simply supported;
(b) all connections of beams, girders or trusses are proportioned to resist the reaction shear forces applied at the appropriate eccentricity;
(c) compression members are subjected to loads applied at appropriate eccentricities (Clause 34) with effective lengths given in Clauses 30, 31;
(d) tension members are designed for loads on the net area. See Clause 42.

In structure designed in accordance with the simple method of design, bracing or shear walls are necessary to provide resistance to horizontal loading; see Section 1.7.5.

1.7.2 Semi-rigid design
Practical joints are not made as perfect pins. They are capable of transmitting some moments which causes a reduction in the sagging moments in beams and additional moments in columns. Economy can be achieved if this partial fixity

is taken into account. Three approaches are offered in BS449 to apply this method of design.

(a) Experimental basis
In general, the design may be based on the experimentally determined behaviour of the joints to be used. The behaviour curve could be incorporated into the analysis.

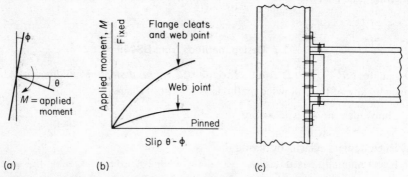

Figure 1.7   (a) Beam—column connection. (b) Joint behaviour curves. (c) Joint—flange cleats and web connections.

A typical behaviour curve and joint is shown in Fig. 1.7. A flange and web joint is also shown. The cleat angles and bolts deform so the full rotation at the end of the beam is not transmitted into the column.

(b) PD3343 — Recommendations
It is stated in BS449 that if the design is made in accordance with PD3343 *Recommendations for design* [8], the requirements of BS449 in respect of stresses may be deemed to be satisfied. This document presents the findings of the Steel Structures Research Committee set up in the 1930s to report on the real behaviour of structures with the rivetted joints then used.

(c) Simplified rules for semi-rigid design
In this method an allowance may be made for the inter-restraint of the connection between a beam and column of 10% of the free moment in the beam when this is assumed to be simply supported. The following conditions must be complied with:

 (i) the beams and columns are designed by the rules for members of a simply supported frame;
(ii) the beams are designed for the maximum net moment with due allowance for any difference in restraint moment at each end;
(iii) each column is designed to resist the sum of the restraint moments from the

beams at the same level on each side of the column in addition to the moments due to eccentricity of connections;

(iv) the end restraint moment need not be taken as 10% of the free moment for all beams, provided the same value is used in the design of the column and beam at each connection;

(v) the column is fully encased in concrete in accordance with Clause 30(b) and the beam-to-column connection includes a top cleat.

Semi-rigid design has never really caught on as a generally used method of design. Designers prefer to use either simple design or fully rigid design. Two reasons for this are:

(a) the general use of computers for analysis has led to greater usage of rigid frame construction;

(b) solid casing of columns for fire protection is not used so frequently as previously and this is a requirement for method (c) above.

1.7.3 Fully rigid design

BS449 states 'This method, as compared with the method for simple and semi-rigid design, will give the greatest rigidity and economy in weight of steel used when applied in appropriate cases'. The main feature is that joints are rigid and the rotation is the same for all members meeting at the joint. Thus rigid design may not give a cheaper building than simple design because the rigid joints are more expensive to fabricate and erection is more costly than for simple design. In a particular case comparative designs and estimates of cost would have to be made to determine the most economical method. However, the decision to use rigid design may be based on other grounds, such as aesthetics and maintenance cost, because the rigid design gives a much cleaner building.

Two methods of rigid design are mentioned in the code.

(a) Elastic theory. Here accurate elastic methods are used for analysis and section design. The members are sized such that permissible stresses given in the code are not exceeded. Using computer analysis large frames can be analysed and designed without difficulty.

(b) Plastic theory. The plastic method is used for the analysis of the structure and the section design. The members are sized so that there is an adequate load factor against collapse. Load factors are not given in BS449 but are taken from the Constrado publication [7]. The code also states that deflections under working loads must not exceed limits set in the code. Plastic design is limited largely to single-storey single- or multi-bay structures. In application to multi-storey structures usually a combination of elastic and plastic design is used.

1.7.4 Experimental basis of design

The code states that where the structure is of such an unconventional nature that calculations are not practicable, load tests shall be made in accordance with procedures set out in Appendix A. The acceptance tests given here are:

(a) Stiffness test. The structure is subjected to the dead load plus a test load of 1½ times the imposed load and this loading is to be maintained for 24 h. If on removal of the load the structure does not show a recovery of 80% of the maximum deflection, the test shall be repeated. The structure is stiff enough if recovery in deflection after the second test is 90% of the increase in deflection during the second test;

(b) Strength test. The test loads in this case are

(i) in general − 2 (dead load + imposed load) in addition to the dead load,
(ii) in dwelling houses of not more than two stories and in schools − 2 x imposed load,
(iii) wind load − 2 x wind loads either with or without the vertical load to give the most severe effect.

The loads shall be maintained for 24 h. The structure has adequate strength if no part completely fails and if on removal of the load the recovery in deflection is 20% of the maximum deflection during the test.

1.7.5 Resistance to horizontal forces and stability

Horizontal loading is due to wind, dynamic loads, seismic loads, etc. Clause 10, BS449, states than when considering the effect of wind load, due allowance can be made for the resistance of and stiffening effects of floors, roofs and walls.

If the floors, roofs or walls cannot carry the loads, then bracing or rigid-frame construction must be provided. Vertical bracing is usually provided in the external walls or around stair wells or lift shafts. External bracing can be made an architectural feature of the finished building. Rigid frames can be used where unobstructed bays for glazing are required.

With regard to stability, the code states that the whole structure or any part of it must be stable. The weight of an anchorage provided shall be such that the least restoring moment including the anchorage shall not be less than

1.2 x overturning moment due to dead loads +

1.4 x overturning moment due to imposed loads.

Wind is treated as an imposed load. Stability must be checked at intermediate stages during construction.

1.8 Design methods from the draft limit state code

The design methods with brief comments are listed below.

(a) Simple design. This is used where the constructions are such that they will not develop moments adversely affecting the structure. Members are designed on the basis that the joints act as pins.

(b) Semi-continuous design. This method is used where the connections provide a predictable degree of interaction between members beyond that allowed for in simple design but less than that provided by full continuity of construction.

(c) Continuous design. The connections provide full continuity. Elastic and plastic methods of design may be used.

(d) Composite design. This method is used where steel structures or members are so connected to other materials that the combined members act as a composite structure or member.

(e) Stressed skin design. 'In this method the cladding is treated as an integral part of the main structure and provides shear diaphragms which are used to resist structural displacement in the plane of loading.' This method is not discussed in this book.

(f) Other means of design. Here loading tests are used to establish the strength and stiffness of the structure.

The above list includes all the methods of design set out in BS449 and, in addition, defines composite design and stressed skin design.

Some brief comments on the draft code are given below.

(a) The draft code places great emphasis on stability of structures. In the commentary to the code, it is stated that one engineer in the design team should be made responsible for stability. The method of providing resistance to horizontal load, i.e. braced frames, rigid construction or shear walls, should be clearly indicated.

(b) A separate section on resistance to accidental damage is included. Key elements are defined as elements the failure of which would lead to failure of the structure. Key elements must be designed to resist accidental damage. Horizontal and vertical tying of the building is specified.

1.9 Structural idealization

This generally means breaking the complete structure down into single elements, trusses or braced or rigid plane frames for analysis and design. It is rarely possible to consider the three-dimensional structure in its entirety. A classification is as follows.

(a) Three-dimensional structures are treated as a series of plane frames in each direction. The division is made by vertical planes. For example, a tower structure may be analysed as a space frame but is more commonly treated as a series of plane frames. See Fig. 1.8(a).

(b) The structure is divided vertically by horizontal planes into roof structure, walls and foundations. These parts are designed separately. The reactions from one part are applied as loads to the next part. The roof structure may be a series of trusses or a complete three-dimensional structure such as a dome. See Fig. 1.8(b).

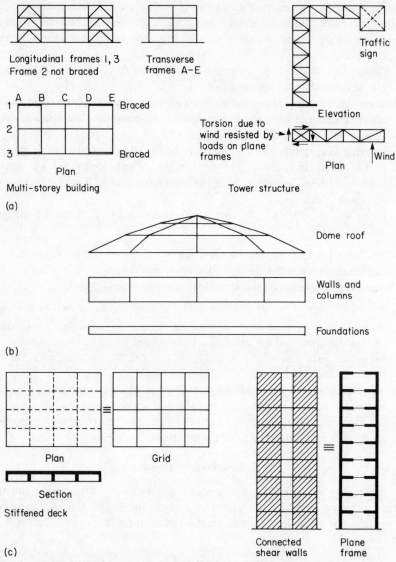

Figure 1.8 (a) Idealization in plane frames. (b) Idealization vertically into separate units. (c) Idealization into a different structural form.

(c) The structure is changed so that the analysis of a different form of structure that is more convenient to carry out, can be made. Examples are (see Fig. 1.8(c)):

(i) a stiffened deck is idealized as a grid;

(ii) a connected shear wall structure is idealized as a plane frame.

Care is needed with the interpretation of results and their use in design. The validity of some idealizations may require proving by experimental work on models.

1.10 Loading

Structures are subjected to dead, imposed, wind and dynamic loads from machinery, earth, materials and fluid pressures, seismic loads, etc. In the design problems treated, actual building loads are used. Some general notes concerning loading are given below.

(a) Dead loads are due to weights of roofs, floors, walls, partitions, finishes and self weights. Typical methods of construction for roofs, floor and walls are specified in the design problems.
(b) Imposed loads. These depend on the intended occupancy or use of the building. These include distributed loads, concentrated loads, impact, inertia, snow, erection and maintenance loads. Imposed loads are given in CP3, Chapter V, Part 1, for various types of buildings.
(c) Wind loads. These depend on the location, the building size and height, openings in walls, etc. Wind loads are given in CP3, Chapter V, Part 2.
(d) Dynamic loads. These are generally caused by cranes where loads are due to vertical impact and horizontal transverse and longitudinal surge. Wheel loads are rolling loads and must be placed in position to give the maximum moments and shears. Dynamic loads for cranes are given in BS449, Chapter 3.

Seismic loads will not be dealt with here. In the problems considered, loading from earth, materials, or fluid pressures do not occur.

1.11 Analysis

The purpose of analysis is to find the design axial loads, shears and moments in all members of the frame. Loads must be applied and separate load cases combined to give the most severe design conditions at various sections. Three methods are used.

(a) Simple design. The structures are statically determinate, and manual analytical or graphical methods of analysis can be used. For very large structures computer programs are needed.
(b) Rigid design – elastic theory. These structures are statically indeterminate and the distribution of moments depends on the member size. Thus the size of members must be estimated before the analysis can be made. Methods used are:

   (i) manual methods such as the use of formulae and charts giving solutions for special cases or the moment distribution method from structural mechanics;

(ii) computer programs such as the I.C.L. plane frame and space frame programs. These use the matrix stiffness method of analysis. The reader should consult text books and structural mechanics, and the computer manuals [12, 13].

(c) Rigid design – plastic theory. The equilibrium method is used for problems considered in the book, where the frame is released and statical and reactant moment diagrams are combined to give the plastic bending moment diagram.

The complete analysis is given for the detailed design cases treated.

1.12 Drawings, specification and quantities

1.12.1 Drawings

Drawings show the arrangement of the structure and detail for fabrication and erection. They are used for taking off the materials list and preparing the bill of quantities and estimates of cost. It is essential that drawings are presented correctly and are accurate. Drawing is an essential part of the design process. The designer must ensure that the detail is such that the structure acts in the way he has idealized it for design. He must also see that all the detailed construction is possible, will not lead to failure and can be painted, inspected and properly maintained.

Steelwork drawings may be classifed as

general arrangement – showing the function and arrangement of the structure
marking plans – showing the location of the separate members
detail drawings – gives details for fabrication.

Many of the consulting engineers design offices carry out the overall analysis and design, only preparing arrangement drawings showing the member sections required. The fabricator then prepares the detail drawings for joints and shop fabrication. This enables him to use details and processes with which he is familiar and has the necessary equipment. Computer detailing programs are now available to produce arrangement and detail drawings, quantities and information for fabrication. Their use will increase in the future.

The present book is mainly concerned with design, and detail drawings are not presented. However, sketches showing framing arrangements, main frames and members and detail for structural joints are given where appropriate. The purpose here is to show the translation of the analysis and design into the practical application.

1.12.2 Specification

The specification and drawings are complementary each providing information necessary for the execution of the work. In general terms the specification includes:

(a) a description of the work to be carried out;
(b) the types and quality of materials to be used;
(c) the standard of workmanship required;
(d) in some cases, the order in which the work is to be carried out and the methods to be used.

Particular clauses in the steelwork specification cover:

(a) the quality of steel required;
(b) the workmanship and fabrication procedures required and the acceptance limits for dimensional accuracy, straightness, drilling, etc;
(c) welding, methods and procedures required to eliminate defects and cracking and reduce distortion. Testing to be carried out and permissible limits for defects;
(d) the types and quality of bolts to be used;
(e) inspection and marking;
(f) erection, giving the tolerance permissible for out of verticality. Procedures for assembly and testing for high-strength friction grip bolt joints and field-welded joints;
(g) fire protection methods to be used for the finished steel frame building;
(h) corrosion protection for exposed steelwork where the surface preparation, protection system and testing required are described.

In all cases the work set out above must comply with the relevant British Standard. The designer must write clauses in the specification to cover special features in the design, fabrication and erection not set out in general clauses in the codes and general conditions of contract. He must ensure that his intentions as to structural action, behaviour of materials, robustness and durability, etc are met.

Experience and great care are needed in writing the specification.

1.12.3 Quantities

Quantities of materials required are taken from the arrangement and detail drawings. Here the materials required for fabrication and erection are listed for ordering. The list comprises the separate types, sizes and quantities of hot- and cold-rolled sections, flats, plates, slabs, rounds and bolts. It is in the general form

> Item number
> Number off
> Description
> Weight per metre/square metre/unit
> Length/area
> Total weight.

The quantities are presented on standard sheets. They may also be printed out from a computer program.

The bill of quantities is a schedule of the materials required and work to be carried out. It provides the basis on which tenders are to be obtained and payment made for work completed. The separate quantities have to be combined and presented under the standard headings in the bill. These are:

Fabrication	Tonnes
Rolled columns	–
Rolled beams	–
Built up columns	–
Trusses, lattice girders	–
Plate girders	–
Purlins and sheeting rails	–
Bracings	–
Bases, grillages, anchorages, H D bolts	–/number

Erection	
Steelwork to be erected	–
Bolts	number

The bill requests the rate and amount for each item. These are then added to give the total cost.

References and further reading

[1] Bressler, B. and Lin, T. T. Y. (1964). *Design of steel structures*. John Wiley, New York.

[2] BS449, Part 2 (1969). *The Use of Structural Steel in Building*. British Standards Institution, London.

[3] *Civil Engineering Standard Method of Measurement* (1976). Institution of Civil Engineers, London.

[4] Document 77/13908 DC (1977). *Draft Standard Specification for the Structural Use of Steelwork*, Part 1, Simple Construction and Continuous Construction. British Standards Institution, London.

[5] Kuzmanovic, B. O. and Willens, N. (1977). *Steel Design for Structural Engineers*. Prentice Hall, Englewood Cliffs N. J.

[6] Lothers, J. E. (1960). *Advanced Design in Structural Steel*. Prentice Hall, Englewood Cliffs N. J.

[7] Morris, L. J. and Randall, A. L. (1975). *Plastic Design*. Constrado, London.

[8] PD3343 (1970). *Recommendations for Design* (Supplement no. 1 to BS449). British Standards Institution, London.

[9] Pippard, A. J. S. and Baker, J. F. (1943). *The Analysis of Engineering Structures*. Edward Arnold, London.

[10] White, R. N., Gergely, P. and Sexsmith, R. C. (1974). *Structural Engineering*, Vol. 3, Behaviour of members and systems. John Wiley, New York.

[11] *The background to the new British Standard for structural steelwork* (1978). Constrado, London.

[12] Ghali, A. and Neville, A. M. (1978). *Structural Analysis*. Chapman and Hall, London.

[13] *I.C.L. Computer Manual* (1969). Analysis of plane frames and grids TP4179. International Computers Limited.

2 Design considerations

2.1 Structural steels

Structural steels are mainly composed of iron with carefully controlled amounts of alloying elements to increase the strength and ensure ductility, impact resistance, weldability, etc. The chemical composition for the various grades of steel is set out in BS4360 *Weldable structural steels* [11]. This gives the following limits:

carbon	0.16–0.25%
manganese	1.5–1.6%
silicon	$\leqslant 0.6\%$.

The phosphorus and sulphur content must not exceed 0.04% because larger amounts adversely affect the ductility. Copper is added to give the weather-resistant steels, Corten, which do not require protection against corrosion.

Structural steels are produced in three grades, 43, 50 and 55, where the numbers are the minimum tensile strengths in hectobars. A grade 50 steel has a minimum tensile strength of 500 N/mm^2. The steels in each grade are further classified according to the ability to resist impact as measured by the Charpy impact test. This classification is shown by the letters A, B, C, D and E, in ascending order of impact resistance. This is discussed under brittle fracture in Section 2.4.

A standard tensile test in accordance with BS18 *Methods for tensile testing of metals* [4] may be carried out and the stress–strain curve drawn. A typical curve for Grade 43 steel is shown in Fig. 1.2. Some of the specified properties for various grades of the structural steels from BS4360 are given in Table 2.1.

Table 2.1   Specified mechanical properties for structural steels

Steel grade	Tensile strength (N/mm^2)	Minimum yield stress (N/mm^2) for thicknesses of (mm)				Minimum percentage elongation on gauge length of 200 mm
		$\leqslant 16$	> 16 $\leqslant 25$	> 25 $\leqslant 40$	> 40 $\leqslant 63$	
43 ABCD	430/510	255	245	240	230	20
43 E		270	260	255	245	20
50 BC	490/620	355	345	345	340	19
55 C	550/700	450	430	415	–	17

Other principal properties of steels which are the same for all steels are:

Young's modulus, E $\qquad = 2.1 \times 10^5 \text{ N/mm}^2$

Poissons ratio $\qquad = 0.3$

Coefficient of thermal expansion $= 12 \times 10^{-5}/^\circ\text{C/unit length}$

Density $\qquad = 157.1 \text{ kN/m}^3$

Weight — section of 100 mm^2 $\qquad = 7.85 \text{ kg/m}$.

The yield strength depends on the thickness of the section. The ductility of steel is very important. This is its ability to deform without fracturing. It decreases with increase in strength.

Structural steel is marketed in hot-rolled plates and shapes as angles, tees, channel and I sections. The strength depends on the chemical composition and the work done on the section. Rolling leaves residual stresses in the section which can have a great effect on the behaviour of the member. These have generally not been taken into account in design. However, the new steel code will do so.

Rolling tolerances are given in BS4 [3] to ensure that sizes of sections and thicknesses of metal are within the prescribed limits. Tolerances for dimensions, thickness and flatness are given in BS4360 [11].

2.2 Jointing

2.2.1 General considerations
Joints are required to:

(a) connect individual frame members together;
(b) join the plates and sections in built-up members to ensure composite action.

Joints are designed to transmit axial load, shear, moment and torsion. In general, a pinned joint transmits axial load and shear while a rigid joint will transmit all actions. Sliding joints to transmit a reaction only are often required where provision for expansion is needed. Joints are made by pins, bolts and welding. Some joint types are shown in Fig. 2.1.

Design and detailing of connections is as important as the frame analysis and design of members. The structure is only as strong as its connections.

2.2.2 Bolting
The two types of bolts in general use in structural steelwork are black bolts and high-strength friction grip bolts (HSFG).

(a) Black bolts
The bolts are forged from round bars with machined threads on bolts and nuts. They are used in holes with 2 mm clearance. The bolts are specified in general

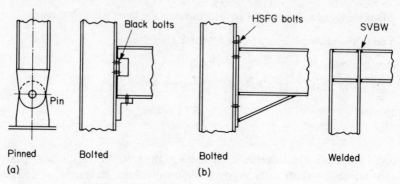

Figure 2.1   (a) Pinned joints. (b) Rigid joints.

purpose joints in simple design and are manufactured in two grades in accordance with BS4190 [10]. The grades are:

Grade 4.6 Mild Steel, yield stress   235 N/mm^2
Grade 8.8 High Strength Steel, yield stress 627 N/mm^2

Bolts may be used in joints in single or double shear or tension or in resisting moment in combined tension and shear or torsion. Examples of these joints are shown in Fig. 2.2. The design of these joints is discussed briefly.

(i) Shear connections. The applied load is transmitted by shear in the bolt shank and bearing on the plate. The safe loads for bolts are given by:

Single shear = nominal shank area x allowable shear stress

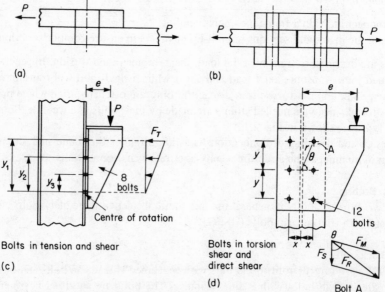

Figure 2.2   (a) Single shear. (b) Double shear. (c) Moment connections.

Double shear = 2 x single shear value

Bearing = nominal diameter x plate thickness x allowable stress in bearing.

Allowable stresses in shear and bearing are given in BS449 [8] and safe load values are given in the BCSA *Structural Steelwork Handbook* [27].

(ii) Tension connections. The bolt is checked on the tensile stress area of the thread. The safe tensile stress is given in BS449 and net areas are given in the BCSA *Structural Steelwork Handbook*.

(iii) Moment connection with bolts in tension and shear (see Fig. 2.2c).

F_T = maximum bolts tension = $Pey_1/[2(y_1^2 + y_2^2 + y_3^2)]$

f_t = maximum tensile strength = F_T/A

A = nominal area of the shank

F_S = bolt shear = $P/8$

f_s = bolt shear stress = F_S/A

p_s = allowable shear stress $\qquad$ BS 449, Table 20

p_t = allowable tensile stress $\qquad$ BS 449, Table 20

$\qquad$ Interaction formula $(f_s/p_s) + (f_t/p_t) \leqslant 1.4$

(iv) Moment connection with bolts in shear due to torsion and direct load (see Fig. 2.2c).

F_M = shear due to torsion

$\qquad = Per/(12x^2 + 8y^2)$

F_S = direct shear = $P/12$

F_R = resultant shear = $[F_M^2 + F_s^2 + 2F_M F_s \cos\theta]^{1/2}$

f_s = shear stress = F_R/A

Examples of joint design are given elsewhere [23, 26].

For detailing, minimum pitches and edge distances are given in Clause 52, BS449 [8] and gauge lines for holes in rolled section are given in the BCSA *Structural Steelwork Handbook* [27].

(b) High-strength friction grip (HSFG) bolts
These bolts are used where

(i) strong joints with bolts in shear or tension are required in simple design;

(ii) connections to transmit large moments as well as shear and axial load are required in rigid design. Grade 8.8 bolts may also be used here.

They are manufactured in two grades in accordance with BS4395, Parts 1 and 2 [12]. For 20 mm diameter HSFG bolts

General grade – proof load   = 144 kN
Higher strength – proof load = 216 kN.

HSFG bolts may be used in all the types of connections shown in Fig. 2.2. The high-strength bolts and nuts must be used with hardened steel washers to' prevent damage to the steel plates when the bolts are tightened. The nuts are screwed up to give a predetermined tension in the bolt shank. The force in the connected members is then transmitted by friction between the surfaces set up by the clamping force. The joint action is different from that in connections with black bolts. The bolts are put in holes with 2 mm clearance.

It is important to ensure that the bolts are tightened to give the required shank tension. Otherwise the joint will act as one with black bolts. Methods used for this are:

(i) torque control uisng a hand torque spanner or calibrated impact wrench to deliver a specified torque;

(ii) part-turning. The nut is tightened and then tightened a further half to three-quarters turn;

(iii) load-indicating washer manufactured with projections which squash to leave a specified gap when the correct shank tension has been reached.

Design for HSFG bolt joints is based on the following principles:

(i) shear connections. The safe load on one bolt in single shear is:

= proof load x slip factor/load factor

where slip factor   = 0.45 for properly prepared surfaces

load factor   = 1.4 for dead and imposed load

= 1.2 with wind load provided the joint is safe for dead and imposed load.

proof load, P_f = minimum shank tension given in BS4395 [12]

double shear value = 2 x single shear value

Safe loads are given in the BCSA *Stuctural Steelwork Handbook* [27];

(ii)   tension connections. The safe load on one bolt

= 0.6 x proof load;

(iii)   moment connections with bolts in tension and shear (see Fig. 2.2c). If the bolt is subjected to an external tension, F_T, then the shear capacity due to friction is calculated using a reduced proof load and is given by:

$(P_f - 1.7 F_T)$ x slip factor/load factor.

The clamping action ceases when the externally applied tension reaches 0.6 of the proof load.

Examples of the design of joints using high-strength friction grip bolts are given elsewhere [23, 26]. Joint design for rigid portals is given in Chapters 5, 6 and 7.

2.2.3 Welding
The use of welding results in the following important advantages:

  (i) the structure is cleaner and better looking;
 (ii) full strength joints can be made giving a lighter structure;
(iii) maintenance costs are lower.

Precautions necessary with welded fabrication are discussed below.

(a) Welding processes
In welding structural steelwork, the parts are fused by heat from the electric arc and jointed together by molten metal from the electrode. The main welding processes are:

  (i) manual arc welding using a hand-held coated electrode. The flux forms a gaseous envelope which protects the arc and the weld pool from interaction with atmospheric gases. It also leaves a slag over the weld which serves as a further protection. Manual welding is slow and the quality of welding depends very much on the skill of the welder;
 (ii) automatic welding, where a continuous wire electrode is fed from a drum through the weld nozzle to the arc where the metal is deposited. The welding machine is carried on a travelling gantry over the work. The girder plates are supported in a frame and are positioned and turned as required for the weld to be made in the down-hand position.

     The electrode may be:

continuously coated with flux,
a bare wire with the flux deposited separately through a tube around the wire — termed submerged arc welding,
a bare wire electrode where a stream of inert gas is directed over the arc to protect it from the atmosphere — termed the metal inert gas (MIG) process. Carbon dioxide is used as the shielding gas.
(iii) Electro-slag welding. This is a process for making full strength butt welds. See Pratt [25] for a description of this process.

(b) Weld detail, cracks and defects
The two main types of weld used are; butt welds and fillet welds. Typical details for welded joints are shown in Fig. 2.3(a). The edge preparation for butt welds is made by machining or flame cutting. Weld edge preparaton is set out in BS5135

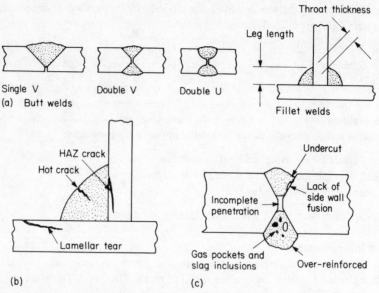

Figure 2.3   (a) Weld detail. (b) Cracking. (c) Weld defects.

Specification for metal arc welding for carbon steel [15].

Cracking can occur in the weld metal or adjacent parent metal. Types of cracking are:

(i) hot cracking caused by contraction of the weld metal;
(ii) heat-affected zone (HAZ) cracking caused by the parent metal becoming brittle. This depends on the rate of cooling, steel composition and the presence of hydrogen;
(iii) lamellar tearing. Contraction of the weld metal causes a tear along a lamination due to a slag inclusion in the plate.

The types of cracking are shown in Fig. 2.3(b). Specialized advice is needed to deal with the problems of hot cracking and HAZ cracking. To reduce the likelihood of lamillar tearing occurring, only sound plates, tested ultrasonically for laminations, should be used.

The main defects which occur in welds are:

(i) over-reinforcement;
(ii) under-cutting;
(iii) incomplete penetration;
(iv) lack of side wall fusion;
(v) gas pockets;
(vi) slag inclusions.

In general, weld defects lead to a reduction in strength of the joint and they may also initiate failure due to brittle fracture or fatigue.

Weld defects are shown in Fig. 2.3(c). BS5135, Appendix E (Avoidance of Hydrogen Cracking) [15] sets out the preheating required for values of combined plate thickness and weld sizes to reduce cracking.

(c) Residual stresses and distortion
When a length of weld cools and solidifies it contracts and sets up residual stresses. These are longitudinal tensile stresses in the weld and compressive stresses in the plate as shown in Fig. 2.4(a). Residual stresses may be of the order of the yield stress. They can be relieved by heat treatment, i.e. stress relieving. However, this is expensive and is not usually carried out with structural steelwork. Account is taken of residual stresses due to fabrication in the new steel code.

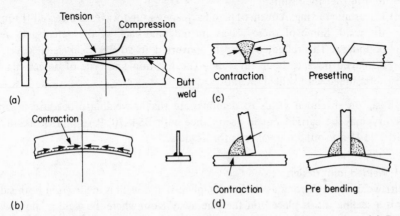

Figure 2.4   (a) Residual stresses. (b) Curvature due to weld cooling. (c) Butt weld. (d) Fillet welds.

In an asymmetrical section, the longitudinal contraction in the weld causes the member to curve as shown in Fig. 2.4(b). Contraction across the weld in an asymmetrical joint will cause distortion. This is shown in Fig. 2.4(c) and (d). In these cases the distortion will be less if double V welds or two fillet welds are used.

Some general guidance may be given to reduce distortion and correct for its effects.

(a) Symmetrical sections with balanced welds will show less distortion.
(b) The sequence of welding used can reduce distortion.
(c) Preheating is necessary with thicker plates to prevent cracking as well as reduce distortion.
(d) Prebending and presetting of plates can be done to offset the effects of distortion due to contraction. See Fig. 2.4(c) and (d).
(e) Heat may be applied after fabrication to correct distortion.

Distortion can be a very severe problem in box girder construction. Expert advice should be sought at the design stage.

(d) Inspection and testing

Inspection is necessary to ensure that the edge preparation, welding procedure and weld sizes are correct. Visual inspection is also made for cracks, undercutting, etc, in the finished weld.

Non-destructive testing is carried out to establish the soundness of welds. The methods used are:

(i) Surface tests. Penetrant dyes and magnetic particles are used to show up surface cracks. The magnetic particle test will also indicate an interior defect;
(ii) X and gamma radiography. The weld is photographed using X or gamma rays. Welds defects are shown up on the film. Care and expertise are needed to interpret the results;
(iii) Ultrasonic testing. A beam of high-frequency sound waves is directed through the weld. Some of these waves are reflected back by the defect and weld boundary. The reflected wave is converted to an electrical signal which is displayed on a cathode ray tube. The location and type of defect can be determined with skilled operation.

Tests on specimen welds to demonstrate that a welding procedure is satisfactory may be carried out in accordance with BS4870, Part 1 [13]. BS4871, Part 1 [14] sets out approval tests for welders.

(e) Welded joint design

Butt welds are as strong as the parent plate if the weld is made from both sides or if a sealing run is placed on the side away from where the weld is laid down. If the weld is made from one side only, BS449 [8] states in Clause 54(b) that the throat thickness should be taken as 7/8 of the thickness of the thinner plate.

For fillet welds the size is specified by the leg length, but the strength is calculated on the throat thickness. For $90°$ fillet welds the throat thickness is 0.7 x leg length. To allow for craters at the ends of fillet welds, BS5135 [15] states that the effective length is equal to the overall length minus 2 x leg length. Strengths of fillet welds are given in the BCSA *Structural Steelwork Handbook* [27].

The design for eccentric connections is summarized below.

(i) Torsion connection. Fig. 2.5(a) shows an asymmetrical weld.

$$L = 2x + y = \text{length of weld}$$

$$\bar{x} = x^2/L$$

$$I_{XX} = y^3/12 + xy^2/2$$

$$I_{YY} = y\bar{x}^2 + x^3/6 + 2x(x/2 - \bar{x})^2$$

$$I_p = I_{XX} + I_{YY}$$

$$r = [(y/2)^2 + (x - \bar{x})^2]^{1/2}$$

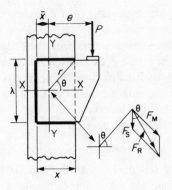 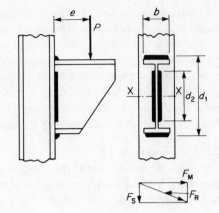

(a)   (b)

Figure 2.5   (a) Torsion connection. (b) Bracket connection.

F_S = direct shear = P/L

F_M = shear due to moment = Per/I_p

F_R = resultant shear per unit length

$\quad = [F_S{}^2 + F_M{}^2 + 2F_S F_M \cos \theta]^{1/2}$

The weld size can be selected from the BCSA *Structural Steelwork Handbook* [27].

(ii) Bracket connection. Fig. 2.5(b) shows the connection. Take the web weld to be half the size of the flange weld.

L = length = $2b + d_2$

I_{XX} = $bd_1{}^2/2 + d_2{}^3/12$

F_M = shear due to moment = $Ped_1/2I_{XX}$

F_S = direct shear = P/L

F_R = resultant shear = $[F_M + F_S{}^2]^{1/2}$

The weld size can be selected from the BCSA *Structural Steelwork Handbook* [27].

2.3 Fatigue

2.3.1 Characteristics of fatigue failure

Fatigue failure occurs in members subjected to variable loads at values significantly below those that would cause failure under static conditions. In static tests to failure there is a large elongation prior to failure, e.g. 20% for mild steel, whereas in fatigue failures there is no plastic elongation.

Structures subjected to fatigue loading include bridges, crane girders, conveyor gantries, mine headframes, offshore oil rig platforms, etc.

Normally the fatigue crack forms at the surface and progresses into the body of the material growing normal to the direction of the cyclic tensile stress by repeated opening and closing of the crack tip. In welded structures the crack may start at a weld defect. The fatigue crack has a smooth surface with markings emanating from the start of the crack. Failure occurs when the crack has grown large enough so that the remaining section cannot carry the load.

2.3.2 Fatigue tests

The usual laboratory test is the rotating bending test made on small cylindrical specimens with or without notches which are loaded as cantilevers or beams. As the specimen rotates the stress at any point varies between equal maximum values of compression and tension. For a given value of maximum stress S the specimen is tested until failure occurs and the number of cycles N at failure is noted. The test is repeated for other values of stress. The relationship between the applied stress and number of cycles to cause failure is found. This is shown in Fig. 2.6(a). The fatigue endurance limit or stress below which the specimen can sustain an infinite number of cycles can be found.

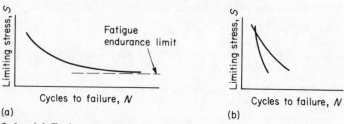

(a)   (b)

Figure 2.6   (a) Endurance limit for specimen from parent material. (b) Fatigue curves for welded joints.

The above is not a great deal of use in the design of welded structural steelwork where stress concentrations and defects in the welds initiate failures. The results of tests on welded joints must be used as a guide. Stresses causing failure for a given number of cycles, i.e. life, are determined for particular welded connections. Typical curves are shown in Fig. 2.6(b). Various types of fluctuating stresses from pulsating tension to alternating tension and compression must be considered.

Some of the types of welded connections tested are shown in Fig. 2.7. The joint giving the least disturbance to the stress flow gives the best result. Some comments on welded joints and welded fabrication are given below.

(a) Butt welds give best performance. Many factors affect the results particularly stress concentrations, abrupt changes of section and weld defects. The fatigue strength is increased by grinding the weld flush.

(b) Joints constructed with fillet welds do not perform well. Joints in order of

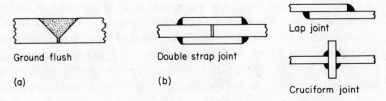

Figure 2.7 (a) Butt weld. (b) Load carrying fillet welds.

performance are shown in Fig. 2.7(b). The effect of non-load carrying fillet welds on fatigue strength is also important.

(c) Girders fabricated with fillet welds between flanges and web have also been tested. The best results are given when the weld is continuous and made by an automatic process. Intermittent welds give low fatigue strength. The effect of stiffeners and splices have also been investigated.

2.3.3 Cumulative damage

Structures are usually subjected to widely varying or random stress cycles. Methods are available for estimating the cumulative damage caused by a given load spectrum. The method is based on the *S–N* curve for the critical weld detail to be used and the limiting stress to give the required number of cycles or life is determined. Alternatively, the life expectancy can be found if a given maximum stress is specified.

2.3.4 Design for fatigue

Design for fatigue is set out in detail in the bridge codes BS153 [7] and the new code BS5400, Part 10 [16]. The method embodies the results of research discussed above and is applicable to design of any member or structure subject to fluctuating load. The main principles are summarized below.

(a) Detail design. All details should be designed to avoid stress concentrations and abrupt changes of section in regions of tensile stress. The thicker plate at a welded splice should be tapered to meet the thinner plate. Splices should be located away from points of maximum tensile stress and should be located preferably at points of contraflexure. Continuous welding using automatic methods should be used in preference to intermittent welding or manual methods.

(b) Working stresses. These are to be reduced to allow for the effects of fatigue. The reduction depends on:

   (i) the value of the maximum tensile stress and ratio maximum stress to minimum stress;

   (ii) the number of stress cycles to which the part is subjected;

   (iii) the constructional detail used.

(c) Constructional detail. The detail is graded in order of ability to resist fatigue failure. No reduction is required for bolted connections. Sound, continuous, full penetration butt welds ground flush give the best performance. Various types of joints with fillet welds are graded according to risk of failure.

(d) Cumulative damage. The design must include an assessment of cumulative damage if the member or joint is subjected to a variety of fluctuating loads.

2.4 Brittle fracture

2.4.1 Characteristic of brittle fracture

Brittle fracture is a low-stress fracture which occurs suddenly with little or no prior deformation. Normally, steel is ductile with an elongation of 16–20%. However, at low ambient temperatures in the presence of other factors, brittle fracture can occur at stresses of one-quarter of the yield stress. The fracture propagates at high speed with the energy required for crack advance coming from stored elastic energy.

The mode of fracture is by cleavage giving a rough surface with chevron markings pointing to the origin of the crack. See Fig. 2.8(a). A ductile fracture occurs by a shear mechanism. The change from shear to cleavage fracture occurs through a transition zone at about 0°C as shown in Fig. 2.8(b). The theoretical study of cracking in steel is made in fracture mechanics.

2.4.2 Notch ductility tests

A number of tests have been devised to determine the notch ductility or resistance of a particular steel to brittle fracture. The most usual test and test recognized in British Standards is the Charpy V-notch test [5]. Here a small beam specimen, 10 mm x 10 mm in section by 55 mm long with a specified notch 2 mm deep at the centre is broken by a striker. The energy in Joules causing fracture is measured. Specimens are tested at various temperatures and fracture energy is plotted against temperature. The transition curve is drawn through the scattered results. See Fig. 2.8(b). The percentage crystallinity of the fracture surface also increases

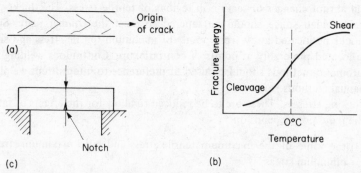

Figure 2.8 (a) Chevron markings on brittle fracture surface. (b) Transition curve. (c) Charpy specimen.

with lowering of temperature as the fracture becomes more brittle. The test is used to determine the amount of energy to fracture a steel at a given temperature. This is the quality control and acceptance criterion for a given steel.

In BS4360 [11] steels are graded A B C D E in order of increasing notch ductility or resistance to brittle fracture. Grade A carries no Charpy test requirement. For a given steel the Charpy V notch impact test fracture energy is specified, e.g. Grade 55C steel is to have an average fracture energy of 27 Joules when tested at $0°$ C. In a location where brittle fracture is likely to occur the steel is specified to have a certain notch ductility.

2.4.3 Factors affecting brittle fracture
Brittle fractures have occurred in a great variety of steel structures but most notably in ships, storage tanks and bridges. One of the best examples of a bridge failure was in the Kings Bridge at Melbourne, Australia, in 1962. Here a plate girder of high-strength steel fractured through. The fracture started in the weld at the end of a cover plate as shown in Fig. 2.9. While the incidence of failure is low, the risk with important structures must not be ignored. The likelihood of brittle fracture is impossible to predict with certainty. Important factors that affect brittle fracture are briefly discussed below.

(a) Temperature. Brittle fractures occur more frequently at low ambient temperatures, i.e. below freezing point.
(b) Stress. Brittle fractures have occurred at low service stresses. However, they are always associated with regions of triaxial tensile stresses.
(c) Plate thickness. The thicker the plate, the higher the risk of brittle fracture occurring. This is because thick sections are more likely to contain defects and complex stress situations.
(d) Welding. Many brittle fractures have started at welds or weld defects but the fracture may propagate through the parent metal. See Fig. 2.9.
(e) Stress concentrations. Brittle fractures often start at stress concentrations due to abrupt changes of section, weld defects, etc.
(f) Type of loading. Brittle fracture has occurred with all types of loading, nominally static loads as well as impact and fatigue loads.

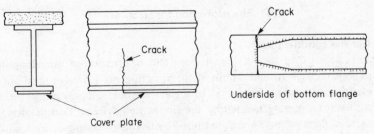

Crack

Crack

Underside of bottom flange

Cover plate

Figure 2.9

(g) Materials. The likelihood of brittle fracture is much higher in some steels. The high-strength steels are more susceptible than mild steel. This leads to the need for notch ductility testing.

2.4.4 Design for brittle fracture

In order to reduce the likelihood of brittle fracture occurring, precautions must be taken in the following areas:

(a) material selection – care and experience are needed to select the material and specify the notch ductility required. More care in design and fabrication is needed with the high-strength steels. Clause 3 in BS449 [8] sets out maximum plate thicknesses for various grades of steel to reduce the likelihood of brittle fracture occurring;

(b) design – great care is needed in design to avoid detail that could lead to brittle fracture. Some points to be noted are:

   (i) abrupt changes of section and stress concentration should be avoided, e.g. the thicker plate must be tapered when it is to be butt welded to a thinner plate,
   (ii) welds in tension should not be placed in regions of high stress,
   (iii) fillet welds should not be made across tension flanges. Cleats to connect secondary members should not be welded to tension flanges. In critical members stiffeners and diaphragms should be bolted to parts in tension,
   (iv) intermittent welding should not be specified,
   (v) the detailing for welds should be such that inspection and testing can be carried out easily.

(c) fabrication – the quality of fabrication is very important. An undetected fault can cause failure. Flame-cut edges should be machined off to remove the heat-affected zone. Welding practice and procedure using preheating for thick plates should be of the highest standard. Complete weld testing should be carried out.

(d) erection – no field practices such as burning holes or tack-welding temporary fixings, etc, should be permitted.

2.5   Fire protection for steelwork

2.5.1 General considerations

Fire causes injury and loss of life, damage and destruction of furnishings and contents and damage to and failure of the structure itself. Design must aim at the prevention or minimization of all of the above effects. Injury and loss of life is caused by toxic gases generated by the fire as well as heat. Loss of property and structural damage and failure are caused by heat and burning of combustible materials.

The means for prevention and control of damage due to fire are classified as follows:

(a) early detection by smoke and heat detectors, followed by extinction of the fire by automatic sprinklers, or manual application of water, foams, etc;
(b) containment by dividing the building into fire-proof compartments to prevent fire spread and smoke travel. Fire-proof escape routes, i.e. stairs and lifts are also necessary;
(c) fire protection of load-bearing structural members to ensure collapse does not occur before people can escape or the fire be extinguished.

The control methods in (b) and (c) form an essential part of the design considerations for steel structures.

2.5.2 Fire and structural steelwork, fire resistance

Structural steelwork performs badly in fires. Temperatures commonly reach 1100–1200° C at the seat of the fire, while the critical temperature for structural steelwork is 550° C. This is the limiting temperature for structural stability when the steel members are designed for stresses in BS449 [8]. At this temperature, the yield point of steel has fallen to about 0.7 of its value at ambient temperature.

Tests are carried out at the Fire Research Station to determine the behaviour of steel elements in fire and also the efficiency of the various methods of protection used.

The ability of a wall, floor, column or beam to remain in position and continue to support load is termed the fire resistance of the structural member. Fire resistance is stated in terms of time, i.e. ½, 1, 2 or 4 h. The fire resistance is determined by testing elements in a furnace where the temperature is controlled to reach 843, 927, 1010 and 1121° C after ½, 1, 2 and 4 h respectively. The fire resistance tests are given in BS476 [6]. The fire resistance needed depends on the type of building, the contents, and the type and location of the structural member. This is given in the Building Regulations [29].

2.5.3 Building regulations

The statutory requirements for fire protection are set out in Part E of the Building Regulations [29]. The following information is summarized from this document with the kind permission of Her Majesty's Stationery Office.

(a) Definitions given in the Regulations.
   Compartment. This is part of a building separated from other parts by compartment floors and walls which are fire resistant.
   Element of structure. This is any member forming part of the structural frame, such as beams and columns or other load-bearing members such as walls and floors.
(b) Purpose groups. Buildings are classified according to use. These groups are related to the fire risk or severity of fire in the building and contents. The

fire risk increases with increase in group number. Some examples of groups defined in the regulations are:

Group I   small residential buildings
Group II   institutional buildings such as schools
Group IV   offices
Group VI   factories

(c) Provision of compartment walls and floors. Buildings must be divided into compartments, the floor area and cubic capacity of which depend on the purpose group and height of the building. The limits are given in a table to the Regulation. For example, for a shop in Purpose group V, there is no height limitation but the floor area of a compartment must not exceed 2000 m^2 nor the cubic capacity exceed 7000 m^3.

(d) Fire resistance of elements of structure. Every element of structure must be so constructed to have a fire resistance not less than the period given in the table to Regulation E5. Periods of fire resistance are given in hours – ½, 1, 1½, 2 and 4 h. The period specified depends on the purpose group, the height, floor area and cubic capacity of building or compartment. Larger values for fire resistance are given for elements in the basement than for elements above ground level. This is because fire is confined in a basement and so its effects can be more severe. Regulation E6 gives the fire resistance periods of floors in conjunction with suspended ceilings. These periods depend on the height of building and type of floor.

(e) Fire resistance periods. The Regulations set out the periods of fire resistance that various forms of construction and fire protection have. These periods have been established by tests at the Fire Research Station. Notional periods of fire resistance are given in Schedule 8 of the Building Regulations. This covers:

Part I: Walls of masonry and framed and composite construction including steel framing.

Part V: Structural steelwork

A – encased steel stanchions – solid and hollow protection
B – encased steel beams – solid and hollow protection

These tables give the type of protection and minimum thickness to give a specified period of fire resistance. See below.

Part VIII: Concrete floors and ceilings required to give a specified period of fire resistance.

A portion of the table in Schedule 8, Part VA on encased steel stanchions is reproduced here in Table 2.2.

2.5.4 Examples of fire protection

Examples of fire protection for columns and floor beams in steel frame buildings are shown in Fig. 2.10.

Table 2.2 Fire protection of steel stanchions

Construction and materials	Minimum thickness (mm) of protection for a fire resistance of (h)				
	4	2	1½	1	½
(A) Solid protection (unplastered)					
1. Concrete not leaner than 1:2:4 mix with natural aggregates					
(a) non-load bearing	50	25	25	25	25
(b) load bearing to BS449	75	50	50	50	50
5. Sprayed vermiculite cement		38	32	19	12.5
(B) Hollow protection					
1. Solid clay bricks	115	50	50	50	50
2. Solid blocks of foamed slag	75	50	50	50	50
3. Metal lathing with gypsum or cement-lime plaster of thickness		38	25	19	12.5

Solid protection by encasement in concrete is not used as much now as in previous times. This type adds considerably to the weight to be carried by the structure and foundations and its cost is not offset by taking its load-bearing capability into account in the case of columns. Various types of hollow protection are preferred for columns. Suspended ceilings may be used to give protection to floor steel. However, ceilings are often not made fire resistant and the protection is applied directly to the steel.

2.5.5 Summary
The procedure to be followed in design is:

(a) classify the building according to its purpose group. Read off the maximum size of compartment;
(b) read off period of fire resistance needed for basement members and members above ground;
(c) select the type of construction and fire protection and the thickness required to provide the necessary fire resistance periods.

Some general concluding notes on fire protection are:

(a) generally all multi-storey steel buildings require fire protection for the structure;
(b) single-storey buildings do not need a fire-resistant structure provided there are adequate means of escape. However, care is needed here particularly if the building is used by small children or elderly people;
(c) multi-storey car parks. Under certain conditions, exposed steel members without fire protection can be used. A relaxation of the Building Regulation must be granted by the local authority;

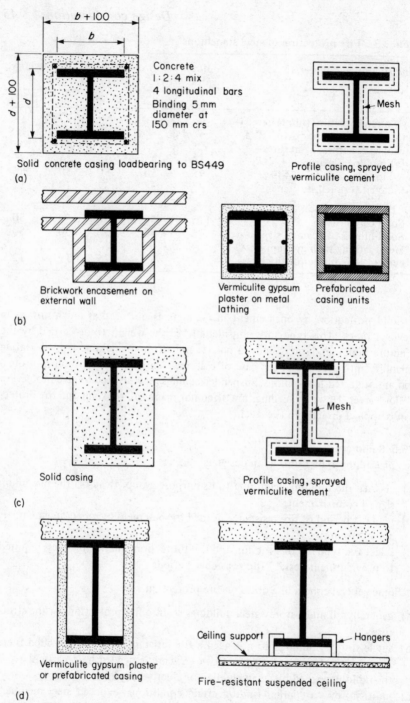

Figure 2.10 (a) Solid casing for stanchions. (b) Hollow casing for stanchions. (c) Solid casing for beams, (d) Hollow casing for beams and suspended ceiling.

(d) in the United States, and some European countries, structural fire protection is provided by using external water-filled box columns. This ensures that the temperature in the steel does not reach a critical level. The external members are constructed in weather-resistant steel.

2.6 Corrosion protection

2.6.1 General considerations
Corrosion is caused by oxidizing of iron in the presence of air and water or other pollutants. This is progressive and leads to failure of the steel member. It is very necessary to provide steelwork with suitable protection against corrosion for the particular environment in which it is located. The choice of the system depends on the type and degree of pollution and length of life required. A long maintenance-free life is possible if the correct system is used and it is correctly applied.

Details regarding all aspects of corrosion protection are set out in BS5493, *Protective coatings of iron and steel structures against corrosion* [17]. A general discussion of the problem and design considerations are given below.

2.6.2 Surface preparation
Surface preparation is the most important single factor in achieving successful protection against corrosion. All hot-rolled steel products are covered with a thin layer of iron oxide which is termed mill scale. If this is not removed, it will break and come off under flexure and abrasion and expose the steel to rusting. All rust, mill scale and slag spatter from welding must be removed before the paint is applied. Methods used in the surface preparation of steel are:

(a) manual cleaning, using scrapers, wire brushes, chip hammers, etc. which is suitable when small areas are involved;
(b) flame cleaning – an oxy-acetylene torch is passed over the surface and the mill scale is loosened by differential expansion. It is then removed by wire brushing;
(c) pickling – The steelwork is immersed in a tank of acid where the oxides are removed by chemical action. This is the method of preparation of steelwork for hot-dip galvanizing. Fabrications up to about 25 mm length can be handled.
(d) blast cleaning – This is the most effective method of removing mill scale and rust. In this method, abrasive particles are directed against the surface of the steelwork at high velocity by compressed air or by centrifugal force from an impeller. The following processes are used.

Open blasting. This is a manual process which can be used on site. Compressed air is generally used to propel the abrasive.
Automatic plant. This is a permanent installation at the fabricators works. The abrasive is thrown by impellers at the finished fabrication.

Abrasives used are chilled iron grit, steel shot, slag, sand, etc.

BS5493 [17] gives standards to assess the quality of the blast-cleaned surface. These are:

cleanliness — First quality. There is no residue of mill scale or rust on the cleaned steel.

Second quality. 95% of the surface is clean.

Third quality. 80% of the surface is clean.

Surface profile — A maximum amplitude of 0.1 mm between peaks and valleys on the surface is recommended.

2.6.3 Types of protective coatings

The protective coatings used are of two main types:

metallic — metal spraying and galvanizing
non-metallic — paint systems.

The coatings are classified in Table 2 of BS5493 [17]. Brief notes are given on some of the main types from this table.

(a) Sprayed metal process. The spraying is carried out by a gun into which metal in the form of a wire or powder is fed. This is fused and atomized by an oxy-acetylene flame and projected by compressed air onto the cleaned metal surface. Zinc and aluminium coatings are used.

(b) Hot-dip galvanizing. Rust and mill scale are removed by pickling in acid. The cleaned metal after fluxing is immersed in a bath of molten zinc. The thickness of coating depends on the time of immersion and the speed of withdrawal.

(c) Paint systems. The paint systems are classified as follows: organic zinc rich, inorganic zinc rich, drying oil types, silicone alkyd, one-pack and two-pack chemical resistant paints and bitumens.

The code gives recommendations on the use of these systems and the thickness of paint film required to give a specified maintenance-free period. Advice should be sought from specialists and manufacturers before specifying the system for a particular application.

Paints are made up of the pigment and binder. The pigment consists of insoluble particles which gives the paint its colour durability and corrosion inhibition. The binder is the non-volatile material that binds the pigment into a film. For drying oil types the main primers used are red lead, zinc phosphate, zinc chromate. Lead-based paints are poisonous and their use may not be permitted in all cases.

The paint system consists of primer, undercoat and finish coat. A blast primer may be applied after blast cleaning and then the main paint system after fabrication. For drying oil types a common undercoat and finish coat is made with micaceous iron oxide as the pigment. The material is in flake-like particles which form a good barrier against corrosive elements.

The bituminous coatings are very resistant to corrosion in sea-water conditions.

2.6.4 Design considerations and specification

The design and detail can greatly influence the life of the coating. Some important points are:

(a) avoid detail that provides places where water can be trapped such as in upturned closed channel sections. Drain holes should be provided if such detail is unavoidable;
(b) use detail that sheds or drains water away;
(c) box sections must be sealed. If properly sealed, BS449 permits the use of thinner sections. Large box sections are often sealed with a desiccating agent inside;
(d) the detail should allow for circulation of air to assist drying of steel after wetting;
(e) Access for future maintenance should be allowed. Where compound sections such as angles or channel back-to-back are used in corrosive atmospheres they should be spaced at 150 mm to allow for painting. In these cases single box sections may be preferable.

BS5493 gives recommendations regarding design and detail.

The code also includes a section on the painting specification. This gives recommendations for clauses on surface preparation, paint systems, control of paint film thickness, application of paint and inspection of the finished coat.

2.7 Fabrication

The general fabrication procedure is set out briefly.

(a) The fabricator prepares the materials lists and drawings showing the shop details.
(b) Rolled members are cut to length and drilled by numerically controlled plant.
(c) Shapes of gussets, cleats, and plates, stiffeners, etc, are marked out with locations for holes and cut and drilled.
(d) For built-up members, plates are flame-cut followed by machining for edges and weld preparation.
(e) Main components and fittings are assembled and positioned and the final welding is carried out usually by an automatic process.
(f) Members are cleaned and painted and given their mark number.

Careful design can reduce fabrication costs. Some points to be considered are:

(a) rationalize the design so that as many similar members as possible are used. This will result in extra material being used but will cut costs;
(b) the simplest detail possible should be used so that welding is reduced to a minimum, good welds can be made and inspection and testing can be carried out easily;
(c) standard bolted connections should be used throughout.

The above is readily achieved in multi-storey building construction.

2.8 Transport and erection

Some brief comments are given regarding the effect of transport and erection on design.

(a) The location of the site and method of transport may govern the largest size of member. Large members may be transported if special arrangements are made.
(b) The method of erection and cranes to be used require careful selection. Economic usage is obtained when building components are of similar size and weight and the crane is used to capacity. Special provisions will have to be made to erect heavy members.

The above considerations effect the number and location of site joints.

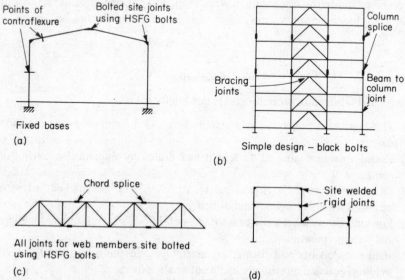

Figure 2.11 (a) Rigid portal. (b) Multi-storey building. (c) All welded rigid frame. (d) Heavy lattice girder.

Factors considered in selecting the type of site joint are: the ease of assembly, appearance and cost. In general, welding is used in shop fabrication and bolting is used for the site joints. Black bolts make the cheapest joints and are used for joints in simple design. High-strength friction grip bolts are used in rigid design and where strong joints are needed in simple design. Site welding is the most expensive form of jointing. It gives the best appearance but the quality may be difficult to control under site conditions. Welding is essential with heavy rigid construction to get full strength joints. Care is needed in design so that the welding and inspection can be carried out easily.

Ideally, joints should be located near points of contraflexure but such positions may not be convenient for fabrication or erection. Site joints on some types of structures are shown in Fig. 2.11.

References and further reading

[1] Boyd, G. M. (ed.) (1970). *Brittle fracture in steel structures*. Butterworths, London.

[2] Bond, G. V. L. (1975). *Water cooled hollow columns*. Constrado, London.

[3] BS4, Part 1 (1972). *Structural steel sections. Hot rolled sections.* British Standards Institution, London.

[4] BS18, Part 2 (1971). *Methods for tensile testing of metals*. British Standards Institution, London.

[5] BS131, Part 2 (1972). *Methods for notched bar tests. Charpy V-notch impact tests on metals*. British Standards Institution, London.

[6] BS476, Part 3, Part 8, (1968–1975). *Fire tests on building materials and structures*. British Standards Institution, London.

[7] BS153, Part 3B and 4 (1972). *Stresses. Design and construction*. British Standards Institution, London.

[8] BS449, Part 2 (1969). *The use of structural steel in building*. British Standards Institution, London.

[9] BS499, Part 2 (1965). *Symbols for welding*. British Standards Institution, London.

[10] BS4190 (1967). *I.S.O. metric black hexagon bolts, screws and nuts*. British Standards Institution, London.

[11] BS4360 (1972). *Weldable structural steels*. British Standards Institution, London.

[12] BS4395 (1969). *High strength friction grip bolts and associated nuts and washers for structural engineering*.
   Part 1, General grade.
   Part 2, Higher grade bolts and nuts and general grade washers.
   British Standards Institution, London.

[13] BS4870, Part 1 (1974). *Approval testing of welding procedure. Fusion welding of steel*. British Standards Institution, London.

[14] BS4871, Part 1 (1974). *Approval testing of welders working to approved testing procedures*. Fusion welding of steel. British Standards Institution, London.

[15] BS5135 (1974). *Metal-arc welding of carbon and carbon manganese steels*. British Standards Institution, London.

[16] BS5400, Part 10 (1980). *Steel, concrete and composite bridge. Fatigue*. British Standards Institution, London.

[17] BS5493 (1977). *Protective coatings of iron and steel structures against corrosion*. British Standards Institution, London.

[18] *Building with steel.* (May 1970). Theme: fire protection. British Steel Corporation, London.

[19] *Building with steel* (February 1975). Theme: fire protection. British Steel Corporation, London.

[20] Chandler, K. A. (1979). Corrosion protection is not all plain painting. *Consulting Engineer*, July.

[21] Elliott, D. A. (1974). *Fire and steel construction. Protection of structural steelwork*. Constrado, London.

[22] Gurney, T. R. (1968). *Fatigue of welded structures*. Cambridge University Press, London.

[23] MacGinley, T. J. (1973). *Structural steelwork calculations and detailing*. Butterworths, London.

[24] *Modern fire protection for structural steelwork* (1967). BCSA, London.

[25] Pratt, J. L. (1979). *Introduction to the welding of structural steelwork*. Constrado, London.

[26] *Steel designers manual*, (1972). Constrado, Crosby Lockwood, London.

[27] *Structural steelwork handbook* (1978). BCSA and Constrado, London.

[28] Welding Institute (1976). *Control of distortion in welded fabrications*. London.

[29] *The Building Regulations* (1976). London, H.M.S.O.

3 Framed buildings—simple design

3.1 Types and general considerations

In framed buildings to simple design, the joints are assumed to be pinned. The assumption gives a conservative design as the action is actually that of semi-rigid design. The structure must be braced to resist lateral loading. However, cantilever action is used in single-storey buildings to give stability. Beams are simply supported but eccentric floor beam reaction give rise to moments in columns. In general, the frameworks are statically determinate with three-dimensional frames idealized into two-dimensional cases for analysis and design.

Any structure may be designed by the simple design method. This method is probably used for the majority of structures. Some of the commoner types of structures may be classified as follows:

Single-storey — single- or multi-bay
Multi-storey — single- or multi-bay.

Examples of braced 'pin jointed' structures are shown in Fig. 3.1.

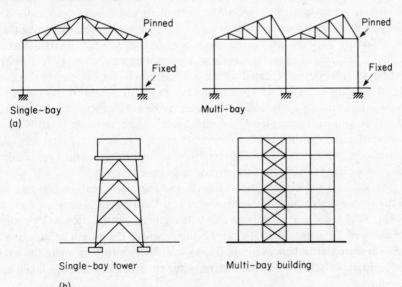

Figure 3.1   (a) Single-storey buildings. (b) Multi-storey structures.

These structures are made from rolled single sections such as universal beams, columns, channels, angles, tees or from compound, laced or battened sections. The structural hollow section used mainly in lattice girder construction also finds considerable application here. This has been the traditional form for steel-framed structures. These structures are cheap and simple to fabricate to normal tolerances and then erect using bolted field joints. The original shop riveting has been entirely replaced by welding. The trussed roof elements occupy dead space, side bracing may interfere with glazing, and maintenance costs are higher than with rigid frame structures.

3.2 Framing plans and typical joint details

Framing plans for a traditional single-storey factory building are shown in Fig. 3.2(a). The columns are fixed at the base and this provides lateral stability against wind loading as shown in the figure. For longitudinal stability, bracing is provided in the end bays in the walls and roof. Bracing and lateral supports are required at eaves level to support the bottom chord of the truss when acting as a strut under uplift wind loads. Variations in the basic arrangement are possible. Fig. 3.2(b) shows the framing for the roof where intermediate trusses are carried on eaves beams. Fig. 3.2(c) shows the effect of knee braces on the transverse action of the frame.

In Fig. 3.3 a flat roof building with columns pinned at the eaves and base is shown. Horizontal, longitudinal wind girders are provided to deliver the transverse wind loads to wind bracing in the gable frames. Note that bracing and longitudinal ties are provided at the level of the bottom chord of the lattice girders. The reactions of these trusses are picked up in the side wall bracing. Framing plans for a factory building with a saw-tooth are given in Chapter 8. As a further example, the framing plans for a multi-storey building are shown in Fig. 3.4. The vertical wind trusses are located on the outside faces of the building in this case. The floor slabs must act as horizontal diaphragms or beams to transmit wind load to the trusses. In general these trusses may be anywhere in the building and are often grouped conveniently around the lift shaft and stair well. Plain or reinforced concrete shear walls or concrete or brickwork infill panels may also be used to provide stability. In Chapter 10 on tall buildings the core-type building using a concrete lift shaft for this purpose is discussed.

The location and classification of the commonly used joints in construction to the simple design method are shown in Fig. 3.5(a). Typical details are shown in Fig. 3.5(b) and (c). In a truss and lattice girder, construction-welded joints are assumed to approximate to pinned joints where the length of members is great compared with their width. In trusses with heavy members where the joints are comparatively large, secondary stresses due to joint fixity must be taken into account.

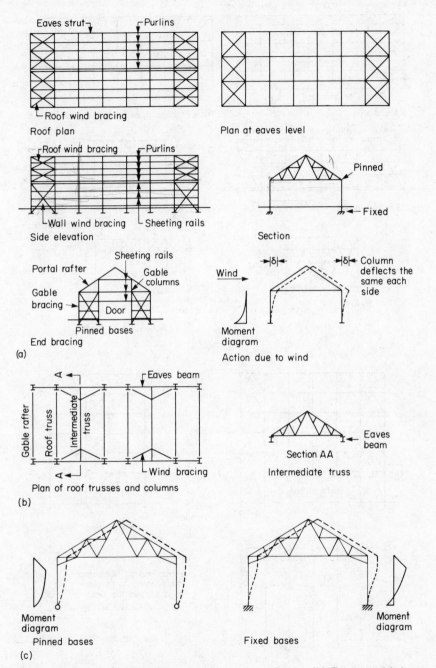

Figure 3.2 (a) Framing for truss and stanchion frame. (b) Frames with intermediate trusses. (c) Frames with knee braces.

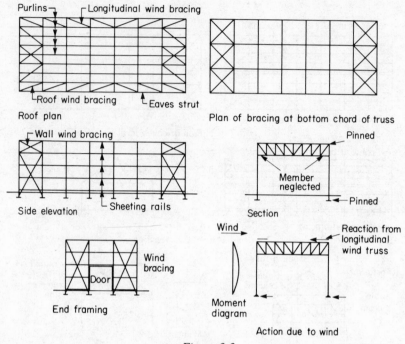

Purlins
Longitudinal wind bracing
Roof wind bracing
Eaves strut
Roof plan

Plan of bracing at bottom chord of truss

Wall wind bracing
Side elevation
Sheeting rails

Pinned
Member neglected
Pinned
Section

Wind bracing
Door
End framing

Wind
Moment diagram

Reaction from longitudinal wind truss
Action due to wind

Figure 3.3

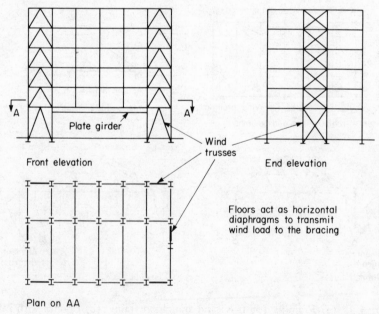

A
Plate girder
A
Front elevation

Wind trusses

End elevation

Floors act as horizontal diaphragms to transmit wind load to the bracing

Plan on AA

Figure 3.4

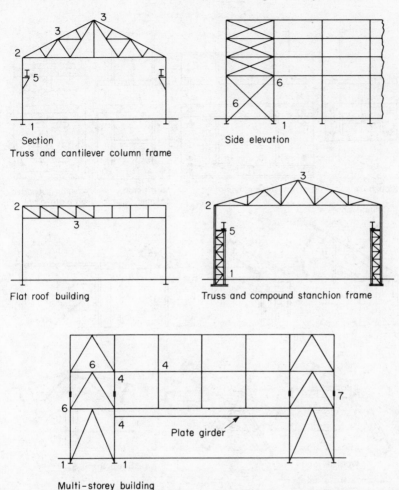

Figure 3.5   Location and classification of joints in typical buildings to simple design. 1 Column bases – pinned and fixed; 2 Truss-to-column joints; 3 Internal truss and lattice girder joints; 4 Beam-to-column and beam-to-beam connections; 5 Crane girder connections; 6 Column splices; 7 Bracing connections.

3.3 Analysis

In general, all members and frames are statically determinate. However, in cases where the structure is redundant either

(a) simplifying assumptions are made so that the analysis can be made by statics, or

(b) methods from statically indeterminate structures are used.

Guidance is also given in BS449 [1] on approximate methods of analysis. Some common cases are discussed below.

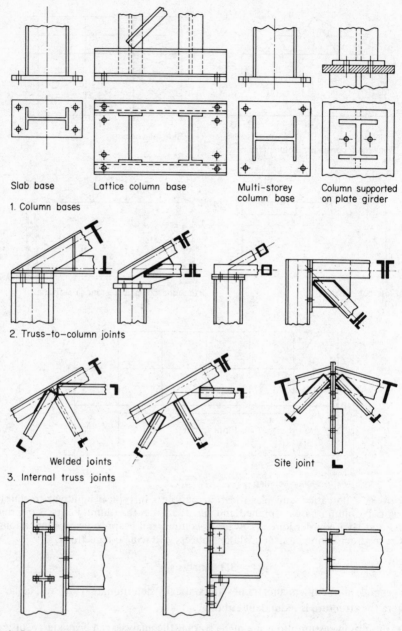

Slab base   Lattice column base   Multi-storey column base   Column supported on plate girder

1. Column bases

2. Truss-to-column joints

Welded joints   Site joint

3. Internal truss joints

4. Beam-to-column and beam-to-beam connections

Figure 3.5(b)

3.3.1 Trusses

Trusses may be analysed by the standard methods: joint resolution, method of sections or the force diagram. In cross-braced trusses, only the tension diagonals

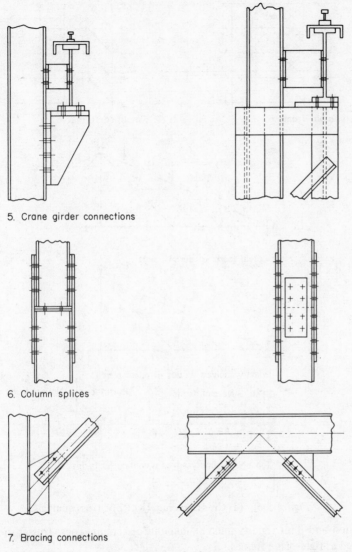

5. Crane girder connections

6. Column splices

7. Bracing connections

Figure 3.5(c)

are considered to be effective. Those in compression are assumed to have buckled and so carry no load. This is shown in Fig. 3.6(a).

In the analysis of roof trusses, if the loads are applied between the panel points, the top chord is analysed separately as a continuous beam supported at the panel points. Alternatively, the top chord can be considered as a set of fixed-ended beams between the panel points. The bending stresses are combined with the primary axial stresses. This is shown in Fig. 3.6(b).

In trusses with short heavy members, secondary bending stresses caused by

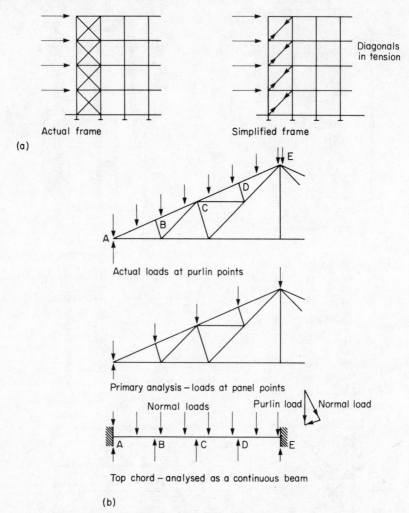

Figure 3.6 (a) Cross bracing. (b) Roof truss analysis.

deflection and joint rigidity must be taken into account. These trusses should be analysed as rigid plane frames.

3.3.2 Frame analysis

The analysis for a truss and column frame is considered first. The frame and loading is shown in Fig. 3.7. The roof truss may be analysed by the force diagram method for the dead, imposed and wind loads shown in Fig. 3.7(b). Reactions from the truss are applied to the columns. The frame as a whole sways sideways under asymmetrical loading. The force P transmitted through the bottom chord can be found by equating deflections at the top of each column. This gives for the cases shown where E is Young's modulus and I the moment of inertia of the

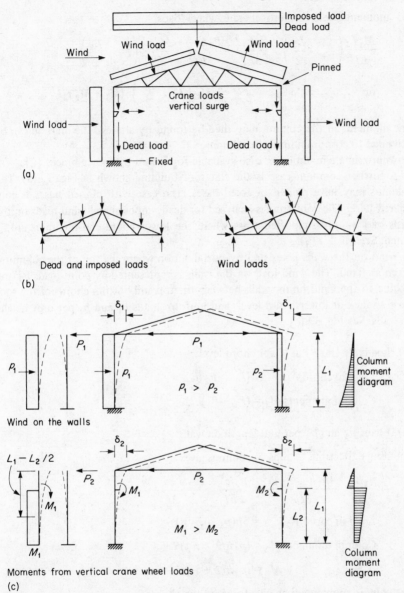

Figure 3.7   (a) Loading on the building frame. (b) Roof truss loads. (c) Loading on columns.

column:

(a) wind on the walls

$$p_1 L_1{}^4/8EI - P_1 L_1{}^3/3EI = p_2 L_1{}^4/8EI + P_1 L_1{}^3/3EI$$

or $\qquad\qquad\qquad P_1 = 3(p_1 - p_2)L_1/16;$

(b) moments from the vertical crane wheel loads

$$\frac{M_1 L_2}{EI}\left(L_1 - \frac{L_2}{2}\right) - \frac{P_2 L_1{}^3}{3EI} = \frac{P_2 L_1{}^3}{3EI} - \frac{M_2 L_2}{EI}\left(L_1 - \frac{L_2}{2}\right)$$

or
$$P_2 = 1.5 L_2\left(L_1 - \frac{L_2}{2}\right)(M_1 + M_2)/L_1{}^3.$$

The moments in the column may then be found by statics. The method can be extended to cover non-uniform columns.

Approximate methods are also available for frames with knee braces [2].

A further common case is the flat roof building shown in Fig. 3.8(a). The columns may have pinned or fixed bases. The case with pinned bases is considered here. The roof truss is analysed for dead, imposed and wind loads on the basis of a simply supported truss where the lower chord end members are ignored. See Fig. 3.8(b).

Wind on the walls is resisted by portal action where the top of the column is taken as fixed. The wind load on the walls is represented by point loads P_1 and P_2 due to the wind on the walls between the top and bottom chords of the roof truss applied at lower chord level, and uniform loads p_1 and p_2 per unit height on walls. See Fig. 3.8(c).

(i) Loads P_1 and P_2 at lower chord level.

Load in chord F_1 $= (P_1 - P_2)/2.$

Column moments $M_1 = (P_1 + P_2)h/2.$

(ii) Loads p_1 and p_2 per unit length on walls.

Equating deflections at tops of columns gives

$$\frac{p_1 h^4}{8EI} - \frac{(p_1 h - F_2)h^3}{3EI} = \frac{p_2 h^4}{8EI} - \frac{(p_2 h + F_2)h^3}{3EI}.$$

Load in chord F_2 $= 5h(p_1 - p_2)/16.$

Column moments $M_2 = (p_1 h/2 - F_2)h$

$$M_3 = (p_2 h/2 + F_2)h.$$

The column moments then give forces in the chords

$$F_3 = (M_1 + M_2)/d$$

$$F_4 = (M_1 + M_3)/d.$$

Hence forces and moments in all members in the frame have been found.

Reference should also be made to Fig. 3.3 where the wind load is carried by longitudinal trusses to bracing in the gable end.

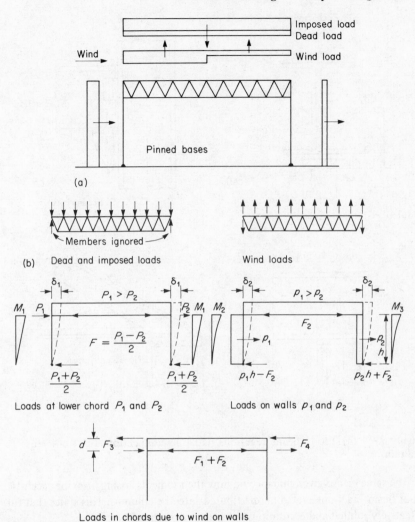

Figure 3.8 (a) Loading on the building frame. (b) Roof truss loads. (c) Wind load on walls.

3.3.3 Eccentricities for floor beams and trusses

Eccentricities for floor beam reactions are given in Clause 34 of BS449 [1]. Here

(a) beam reactions are taken at 100 mm from the face of the section or at the centre of the bearing if this is greater;
(b) in cap connections, the reaction is taken at the edge of the column or packing;
(c) for roof trusses the load is taken as axial. These cases are shown in Fig. 3.9.

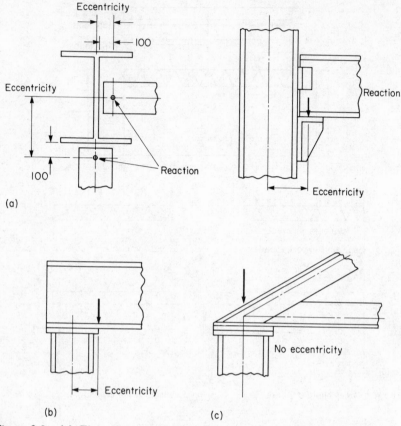

Figure 3.9 (a) Floor beam reactions. (b) Cap connection. (c) Roof truss to column.

The same clause gives guidance on how the moments arising from the eccentric floor beam reactions are to be distributed into the columns. This states that for effectively jointed and continuous columns the moments are:

(a) ineffective at floors above or below the floor considered;
(b) divided equally between the column lengths above and below that floor, provided the stiffness of one length does not exceed 1.5 times that of the other length. If it does, the moment is to be divided in proportion to the stiffnesses. Here:

Stiffness = moment of inertia/length = I/l.

3.4 Design

Important principles involved in the design of structural elements are discussed briefly.

3.4.1 Tension members

Tension members occur in trusses, bracings and hangers. Rounds, rolled sections, cables, etc, are used. The design is made for stress on the net section; see Fig. 3.10. If an angle is connected through one leg the outstanding leg is not fully effective. The net section is specified in Clause 42 of BS449.

$$\text{Net area} = a_1 + a_2\left[3a_1/(3a_1 + a_2)\right],$$

where a_1 is the net area of connected leg, and a_2 is the area of unconnected leg. The complete net area can be used for double angles placed back to back and connected to each side of a gusset.

Safe loads are given in the B.C.S.A. *Structural Steelwork Handbook* [3] .

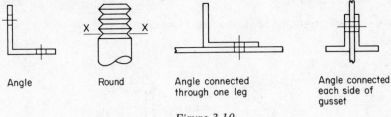

Angle   Round   Angle connected   Angle connected
                                     through one leg   each side of
                                                             gusset

Figure 3.10

3.4.2 Compression members

These members form columns, truss members and bracings. Rolled, compound and build-up sections are used; see Fig. 3.11. Failure, except in very short members, occurs by buckling. The safe load depends on the slenderness ratio as well as the area of cross-section where:

$$\text{Slenderness ratio} = \frac{\text{effective length}}{\text{least radius of gyration}} = \frac{l}{r}.$$

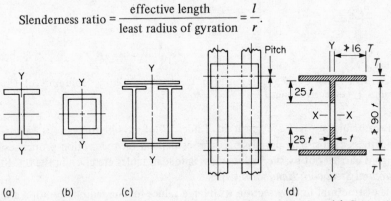

(a)   (b)   (c)   (d)

Figure 3.11   (a) Universal column. (b) Structural hollow section. (c) Compound column. (d) Built up column.

The effective length is:

(a) less than or equal to the actual length in braced construction where the ends of the member are held in position;

(b) greater than the actual length in unbraced construction where the ends can move relative to each other.

Values of effective lengths may be estimated from Clause 31 or taken from values given for various practical cases in Appendix D of BS449. Some cases are shown in Fig. 3.12.

Maximum values for slenderness ratios are given in Clauses 33 and 44 of BS449. These are:

A member carrying dead and imposed load	180
A member carrying wind load	250
A member normally in tension but subject to reversal of stress due to wind	350.

The values may govern the minimum sizes of members.

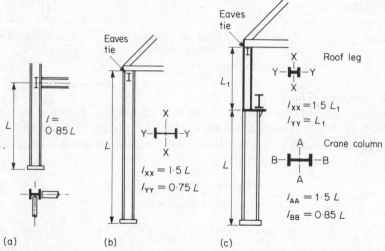

Figure 3.12 (a) Corner column. (b) Cantilever building column. (c) Crane column. l = effective length.

Permissible stresses are given in Table 17 of BS449 for various values of slenderness ratios. The permissible stresses depend on the section properties so design is carried out by successive trials. Safe load tables are given in the B.C.S.A. *Structural Steelwork Handbook* [3].

The structural hollow section with large values for the radii of gyration about each axis is more efficient as a section than the universal column which is weak with respect to buckling about the YY axis. See Fig. 3.11(a) and (b). These sections find a wide use in lattice girder and truss construction and for single bracing systems in industrial buildings.

It is possible to arrange the spacing of members in a compound column so that the column has equal strength with respect to each axis. The pitch for the lacing or battens must be such that the individual member does not fail between the supports. The requirements for design are given in Clauses 35 and 36 of BS449. See Fig. 3.11(c).

For built-up columns made from unstiffened plates from Grade 43 steel, the following rules from Clause 32 of BS449 apply:

(i) the maximum outstand of compression flange must not exceed 16 times the flange thickness;
(ii) the web length must not exceed 90 times its thickness and only 50 times its thickness is taken to be effective in resisting compression. This reduced area is used for calculating the radius of gyration. The gross area is used for all other properties.

A built-up column is shown in Fig. 3.11(d).

3.4.3 Beams

Beams resist load in bending and shear. Any section may serve as a beam but the typical beam section is the universal beam. If the universal beam is not strong enough or in cases where the depth is limited, cover plates are added to give the compound beam. See Fig. 3.13(a) and (b).

If the compression flange is fully restrained, e.g. by a floor slab the beam is sized for

(a) strength $Z_{XX} = M/p_{bc}$ = modulus of section

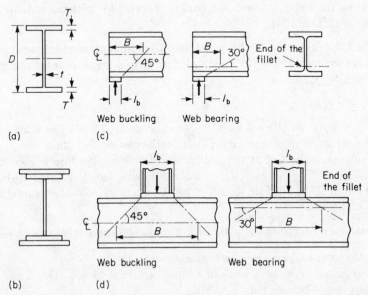

Figure 3.13 (a) Beam section. (b) Compound beam. (c) End bearing. (d) Intermediate bearing.

where p_{bc} is the allowable compressive stress (Table 2 of BS449) and M is the maximum bending moment;

(b) deflection. The code requirements are given in Clause 15 of BS449. Here the deflection due to loads other than dead loads is limited to span/360. For most designs it is advisable to limit the deflection due to the total loads to span/360. For uniformly distributed loading this gives

$$I_{XX} = \text{moment of inertia}$$

$$= 2.34 \ WL^2 \ \text{cm}^4$$

where W is the total distributed load (kN), and L is the span (m).

The beam size may be selected from tables in the B.C.S.A. *Structural Steelwork Handbook* [3] to give the required values of Z_{XX} and I_{XX}.

If the compression flange is not restrained laterally, the allowable stress depends on values of l/r_{YY} and D/T where l is the effective length, (See Clause 26 of BS449), r_{YY} is the radius of gyration about the YY axis, D the overall depth of the beam, and T the thickness of the compression flange. Values of permissible stresses are given in Table 3 of BS 449 [1].

The beam is checked for shear. The average shear stress is

$$f_q' = V/Dt$$

where V is shear, and t web thickness. Values of allowable average shear stress are given in Table 11 of BS449.

Depending on the type of support used and whether concentrated loads are applied to the beam, the web will require checking for buckling and crushing. See Fig. 3.13(c) and (d).

(a) Buckling is checked at the centre line. The load is assumed to spread at 45° from the ends of the stiff bearing length l_b to a length B at the centreline. The safe load

$$W = p_c Bt$$

where p_c is the allowable axial stress for struts, Table 17 of BS449 for a slenderness of $d\sqrt{3}/t$, and d the clear depth between root fillets.
(b) Crushing is checked at the end of the root fillet. The load is assumed to spread at 30° from the ends of the stiff bearing length l_b to a length B at the root fillet. The safe load $W = 190 \ Bt$ where the allowable bearing stress is 190 N/mm².

Safe load values are given in the B.C.S.A. *Structural Steelwork Handbook* [3].

3.4.4 Members subjected to axial load and moment

The interaction formula is given in Clause 14 of BS449 [1]. This states that members should be such that

$$f_c/p_c + f_{bc}/p_{bc} \leqslant 1$$

where f_c is the calculated average axial compressive stress, p_c the allowable compressive stress from Table 17 of BS449, f_{bc} the resultant compressive stress due to bending about both axes, and p_{bc} the allowable compressive stress for members subjected to bending (Clause 19).

Refer also to the rigorous method given in Section 7.3.4.

3.4.5 Plate girders

Plate girders with vertically stiffened webs are considered as shown in Fig. 3.14(a). The following summarizes the steps in design.

(a) For girders with fully restrained compression flanges, the allowable stresses p_{bc}, p_{bt} from Table 2 of BS449 depending on the plate thickness apply to both the tension and compression flange. For a given depth D, the flanges

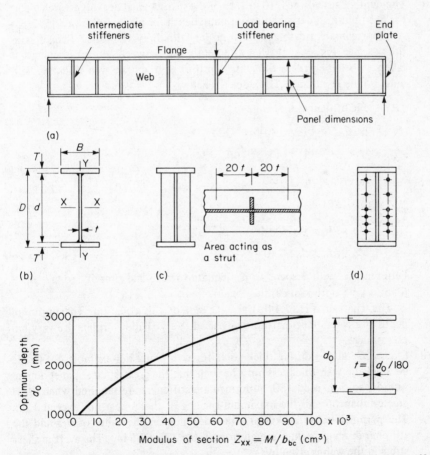

Figure 3.14   (a) Girder elevation. (b) Section. (c) Section at load-bearing stiffener. (d) End plate. (e) Optimum depth chart.

can be sized to resist the total moment M:

Flange area $A_f = M/p_{bc}D$ approximately.

A suitable flange section can be selected. The maximum outstand $(B - t)/2$ for flange plates is given in Table 14 of BS449. For Grade 43 steel the maximum outstand for the compression flange is $16T$ and the tension flange is $20T$, where T is the flange thickness. The girder section is shown in Fig. 3.14(b). Final stresses are checked on the complete girder section after the web has been designed to meet requirements set out in (d) below.

(b) If the compression flange is not fully restrained the allowable compressive stress depends on the girder section and must be determined as set out in Clause 20 of BS449. A method of successive trials is necessary to determine the girder section.

(c) The depth is usually at 1/10 to 1/15 of the span. A great deal of research effort has been put into deriving the optimum depth for plate girders, see [5,7,8]. An approximate treatment for a girder with fully restrained flanges is as follows (see Fig. 3.14e). Here the depth, d_o, between centres of flanges is taken as being equal to the clear depth of web, d. The web thicknesses, t, must not be less than $d_o/180$. See (d) below.

A_f = area of flange

$I_{XX} = 2A_f d_o^2/4 + (d_o/180)(d_o^3/12)$

$Z_{XX} = 2I_{XX}/d_o = A_f d_o + d_o^3/1080 = M/p_{bc}$

$A_f = Z_{XX}/d_o - d_o^2/1080$

A = total area

$\quad = 2Z_{XX}/d_o - d_o^2/540 + d_o^2/180$

$\quad = 2Z_{XX}/d_o + d_o^2/270$

Differentiate with respect to d_o, equate to zero and simplify to give $d_o = 6.46Z_{XX}^{1/3}$ optimum depth.

The chart in Fig. 3.14(e) gives a plot of this equation. The optimum depth is greater than that normally used. Very shallow girders are very uneconomical.

(d) Clause 27e gives limiting thicknesses for web plate. For example, for Grade 43 steel the web thickness must be such that the greatest clear panel dimension does not exceed $180t$. Intermediate stiffeners are required when d/t is greater than 85 and the maximum spacing of these must not exceed $1.5d$. The permissible average shear stress depends on the value of d/t and the stiffener spacing. Values are given in Table 12 of BS449. The average shear stress in the web is given by

$f'_q = V/dt$

where V is the shear force at the section.

(e) Intermediate stiffeners are provided to prevent the web buckling due to shear. The maximum spacing is given in (d) above. Code requirements are given in Clause 27(b). If flats are used the maximum outstand is not to exceed $12t$ where t is the thickness of flat.

(f) Load-bearing stiffeners are required at all load points and supports. These may be designed as struts with an effective length of $0.7d$ where the strut section consists of the stiffeners and 20 times the web thickness on each side of the centre line of the stiffeners. Requirements are given in Clause 28a. See Fig. 3.14(c).

(g) The web flange weld is designed for the horizontal shear calculated by the formula

$$\text{Shear} = VBT(D - T)/4I_{XX} \text{ per weld per unit length}$$

where I_{XX} is the moment of the girder about the XX axis. The weld may be selected from the B.C.S.A. *Structural Steelwork Handbook* [3].

(h) Deflection may control the design in some cases. This is calculated by normal methods.

3.4.6 Crane girders

These members have to resist vertical and horizontal loads from cranes as set out in Clause 7 of BS449[1]. The loads are:

(a) Vertical loads — 25% to be added to the maximum static loads to allow for impact

(b) Horizontal loads transverse to the rails — 10% of the combined weight of the hook load and weight of the crab. This load is to be divided between all the crane wheels.

(c) Horizontal load acting along the rails — 5% of the static wheel loads.

Permissible stresses may be increased by 10% where loads (b) or (c) are taken as acting with load (a).

The loads must be so placed in position to give the maximum bending moment and shear force. This is shown in Fig. 3.15(a).

Consider the light crane girder shown in Fig. 3.15(b). Properties for various combinations of universal beams and channels are listed in the B.C.S.A. *Structural Steelwork Handbook* [3]. The allowable compressive stress depends on l/r_y and D/T.

l = effective length of compression flange, i.e. the span for a simply supported girder

r_Y = radius of gyration of the whole girder about YY axis

D = depth of girder

T = mean thickness of flange = horizontal area A/width W.

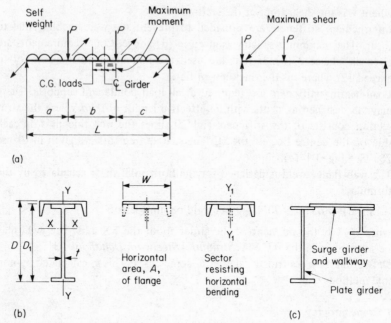

Figure 3.15   (a) Maximum moment and shear. (b) Light crane girder. (c) Heavy crane girder.

Values of allowable stresses are given in Table 3 of BS449. Values for the allowable tensile stress are given in Table 2 of BS449. The section resisting horizontal bending is taken as the channel and top flange of the universal beam. The bending stresses are:

Top flange $f_{bc} = M_{XX}/Z_T + M_H/Z_H$

Bottom flange $f_{bt} = M_{XX}/Z_B$

where M_{XX} is the vertical bending moment, M_H the horizontal bending moment, Z_T the modulus of the top flange for vertical bending, Z_B the modulus for the bottom flange for vertical bending, and Z_H the modulus for section resisting horizontal loads. The average shear stress $f'_q = V/D_1 t$, where V is the maximum shear, and D_1 is the depth of the universal beam. Values of allowable average shear stress are given in Table 11 of BS449.

The vertical deflection is not to exceed span/500. The deflection at centre is very close to the maximum deflection. This is given by:

$$\delta = \frac{PL^3}{48EI}\left[\frac{3(a+c)}{L} - \frac{4}{L^3}(a^3 + c^3)\right] + \frac{5WL^3}{384EI}$$

where P is the vertical crane wheel load, and W the self weight of girder.

3.4.7 Purlins and sheeting rails

(a) Purlins
Purlins carry roof sheeting. Angle and tubular purlins may be designed in accordance with Clause 45 of BS449 [1]. The provisions regarding lateral stability, Clause 26, and limiting deflections of beams do not apply to purlins. The design for angle purlins for roof slopes not exceeding 30° is as follows: L is the span of purlin (mm), and W the distributed load (kN). This must include a minimum imposed load of 0.75 kN/m^2.

The leg length in plane of load $\not< L/45$

The other leg length $\not< L/60$

The modulus of section $\not< (WL/1.8)10^{-3}$ cm^3

Alternatively, cold-rolled purlins can be used. Various types are available and sizes may be selected from manufacturers catalogues [6].

(b) Sheeting rails
These are to be designed in accordance with Clause 46 of BS449 [1] and are to carry horizontal wind loads and vertical loads if required. Limitations regarding deflections and lateral stability of beams do not apply.

Various types of cold-rolled sheeting rails are available and manufacturers literature should be consulted to select sizes [6].

3.4.8 Column bases
The base spreads the column load to deliver it at a safe intensity to the concrete. The main types are slab, gusseted, pocket and grillage. Brief notes are given on slab and pocket bases. See Fig. 3.16.

The size of the base depends on the safe bearing pressure of the concrete. Safe values for bearing pressures p_b are

Mix	Cube strength (N/mm^2)	Safe bearing stress (N/mm^2)
1 : 1 : 2	30	5.4
1 : 1½ : 3	25.5	4.7
1 : 2 : 4	21	4.2

(a) Slab base axially loaded. See Fig. 3.16(a).

Area required: $b \times l = W/p_b$.

W = column load

The slab thickness is given in Clause 38 of BS449 [1]. Here

$$t = \left[\frac{3w}{p_{bct}} \left(A^2 - \frac{B^2}{4} \right) \right]^{1/2}$$

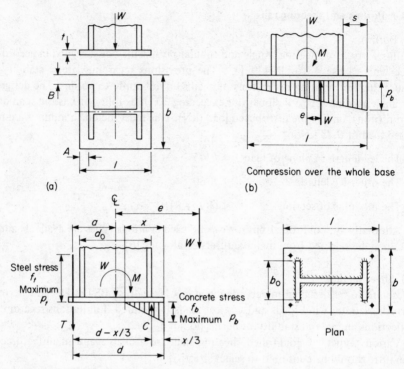

(a)

(b)

Compression over the whole base

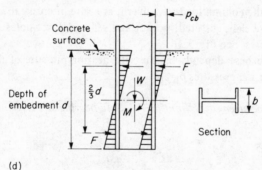

Compression on part of the base — Tension in the holding down bolts
(c)

(d)

Figure 3.16   (a) Slab base – axial load. (b) Slab base – load and moment. (c) Slab base – load and moment. (d) Pocket base.

w = the pressure under base = W/bl N/mm^2

p_{bct} = the permissible stress in bending in steel = 185 N/mm^2

A = the greater projection of the base (mm)

B = the lesser projection of the base (mm).

The column and base should be machined so that the load is transmitted in bearing. Nominal welding is required to hold the base slab in position.

(b) Slab base subjected to an axial load W and moment M. The case where compression occurs over the whole of the base is considered (see Fig. 3.16(b)). This case occurs where the eccentricity $e = M/W \leqslant l/6$

Base area A = width b x length l

Modulus $Z = bl^2/6$

Maximum pressure $p_{max} = W/A + M/Z \leqslant p_b$.

Successive trials are necessary to determine the base size. The column and slab should be machined and nominal welding only is required. The base plate thickness is determined by designing the slab to cantilever a distance s out from the column at the heaviest loaded end.

(c) Slab base subjected to an axial load W and moment M. In this case compression occurs over part of the base and tension in the holding down bolts (see Fig. 3.16(c)). The base is designed using elastic theory for reinforced concrete beam design.

p_t = allowable tensile stress in the holding down bolts

p_b = allowable bearing pressure on the concrete

m = modular ratio = $\dfrac{\text{Young's modulus for steel}}{\text{Young's modulus for concrete}}$ = 15

d = dimension from centre of bolts to the edge in compression

x = position of the neutral axis = $15 p_b d/(15 p_b + p_t)$

$M' = M + Wa$ = moment about centre of the bolts

C = force in compression = $M'/(d - x/3)$

p_{max} = maximum pressure = $2C/bx$

T = tension in bolts = $C - W$

f_t = stress in bolts = $T/(A_{net}N)$

A_{net} = tensile stress area of the bolt

N = number of bolts

The weld from the column to the base needs careful design. A conservative value for the maximum weld load in the flange is given by

$$\text{load} = \frac{M}{b_o d_o} - \frac{W}{2 b_o}$$

where b_o is the width of flange of column, and d_o the distance centre-to-centre of column flanges. The weld type, either fillet, partial butt or butt, may be selected to give the strength required.

The base plate thickness is determined by designing the slab to cantilever from the column either under pressure from the concrete at one end or from tension in the bolts at the other end.

It may be necessary to analyse a given base using elastic theory. The base can be considered to be loaded with W at eccentricity e where, see Fig. 3.16(c)

$e = W/M$.

Define the following terms:

f_b = compressive stress in the concrete

f_t = tensile stress in the bolts = $15f_b(d-x)/x$

C = compressive force in the concrete = $\frac{1}{2}f_b bx$

T = tensile force in the bolts = $f_t A_{net}$

W = axial load on the base = $C - T$

M = moment on the base = We

$$= C[(l/2) - (x/3)] + Ta \tag{3.1}$$
$$= (C - T)e \tag{3.2}$$

Equating the two equations for M gives a cubic equation which can be solved for x. Then the stresses f_b and f_t can be found by back substitution.

(d) Pocket base (see Fig. 3.16d). The column is grouted into a pocket cast in the foundation. The axial load is resisted by bearing and bond between the column and concrete. The moment is resisted by compressive forces acting on the flanges. Define the terms:

p_{cb} = permissible stress in concrete in bending

p_{bt} = permissible bending stress in the steel

Z_{XX} = modulus of section of the column

d = depth of embedment of the steel column

b = breadth of column flange.

Then the moment of resistance of the concrete is equated to the strength of steel column in bending to give the depth of the embedment

$$d = [3p_{bt}Z_{XX}/(p_{cb}b)]^{1/2}.$$

3.5 Coursework exercises

1. A large sign 10 m long by 3 m high is to be supported such that the lower

edge is 5 m above ground level. Draw the framing plans for the supporting
structure.

2. A water tank 6 m x 6 m x 3 m high is to be supported over a factory building.
   The supporting frame is to be independent of the existing building as shown
   in Fig. 3.17.

   (a) Draw the framing plans for the supporting structure.
   (b) Indicate the design loading on sketches.
   (c) List the steps in the analysis and design process.

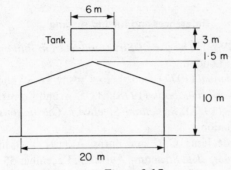

Figure 3.17

3. Draw the framing plans including stairways for a footbridge to cross a road.
   The clear span is 12 m and the clear height from road to underside of bridge
   is to be 6 m.
4. A restaurant is to be built over a motorway. The locations for the foundations
   and clearances required are shown in Fig. 3.18. The imposed load on the

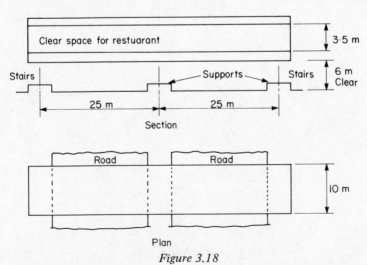

Figure 3.18

restaurant floor is 6 kN/m^2 over the whole floor area.

(a) Draw the framing plans showing the steel required for supports steel frame and stairways. Indicate clearly the method of resisting wind loads.
(b) Estimate the loading due to dead imposed and wind loads and show these loads on sketches.
(c) Make preliminary design calculations to establish the sizes of the main members and foundations. Take the safe bearing pressure to be 150 kN/m^2.

References and further reading

[1] BS449, Part 2 (1969). *The use of Structural Steel in Building*. British Standards Institution, London.
[2] *Steel Designers Manual* (1972). Constrado, Crosby Lockwood, London.
[3] *Structural Steelwork Handbook* (1978). B.C.S.A. and Constrado, London.
[4] MacGinley, T. J. (1973). *Structural Steelwork Calculations and Detailing*. Butterworths, London.
[5] Jenkins, W. M., de Jesus, G. C. and Burns, A. (1977). Optimum Design of Welded Plate Girders. *The Structural Engineer*, December **55** (12).
[6] Ward Brothers Sherburn Ltd. (1979). *Technical Data Cold Rolled Purlins*. Cold Rolled Sheeting Rails.
[7] *Autofab Handbook* (1978). Redpath Dorman Long Limited, Middlesbrough.
[8] Knowles, P. (1977). *Design of Structural Steelwork*. Surrey University Press, London.

4 Simple design—examples

4.1 Introduction

Two examples of traditional structures designed in accordance with the method of simple design set out in Clause 9 of BS449 [1] are presented to show the application of the method. These are:

(a) single-storey factory building;
(b) two-storey office and warehouse building.

The single-storey factory building consists of truss and cantilever crane columns. The design includes:

(a) roof truss analysis and design;
(b) crane girder design;
(c) analysis of the building frame;
(d) design of a crane column;
(e) design of a gable column;
(f) design of the bracing.

Sketches and an arrangement drawing are included to present the design output.

The two-storey building is to be erected over existing premises to permit operations to continue uninterrupted. The structural solution adopted is to suspend the office floor from the roof truss. The building is fully braced. The design includes:

(a) framing plans for the structure;
(b) design of roof and floor steel;
(c) design of a hanger;
(d) design of a roof truss;
(e) design of a side column.

Design of the gable columns and bracing is not included. The main structural details are shown on sketches.

4.2 Simple design — single-storey factory building

4.2.1 Specification

The framing plans for a single-storey single-bay factory building with crane are shown in Fig. 4.1. The frames are at 6 m centres and the length of the building is 60 m. The crane span is 18.0 m.

The following design work is to be carried out using Grade 43 steel:

(a) design the roof truss using structural hollow sections;
(b) design the crane girder;
(c) design the crane column;

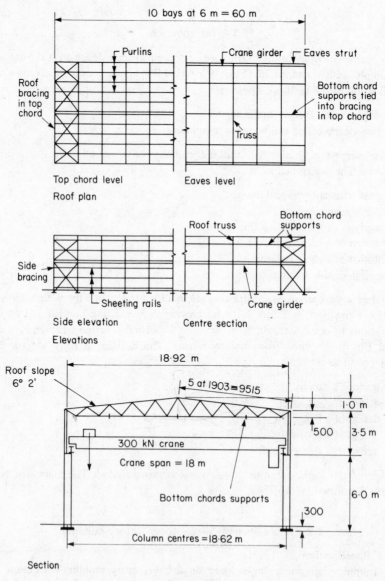

Figure 4.1

(d) design a gable column;

(e) design the bracing.

4.2.2 Loading

Brief details are given for the sheeting and purlins to be used:

Sheeting   Cellactite 11/3 corrugated sheeting, type 575 thickness 0.575 mm

Maximum purlin spacing = 2.0 m

Dead load = 0.092 kN/m^2

Roof pitch 6°2′. End lap 225 mm with flexible lap sealing strips

Purlins   Load — sheeting, insulation, imposed load = 1.0 kN/m^2

Span 6 m, spacing 2 m

Ward Bros. Purlin 80069 Safe Load = 1.13 kN/m^2

The design loads for the building are as follows:

Roof dead load	(kN/m^2)
Sheeting	0.1
Insulation	0.15
Purlins — 5.312 kg/m	0.03
Truss and bracing	0.09
Total load on slope	0.37

Roof imposed load on plan = 0.75 kN/m^2

Walls — sheeting insulation, rails, bracing = 0.3 kN/m^2

Stanchion — estimate loads using assumed sections

Roof leg: 305 x 165 UB 40 = 1.5 kN

Crane portion: 610 x 229 UB 140 = 9.0 kN

Crane girder 610 x 229 UB 140)
$$\left.\begin{array}{l} 381 \times 102] 55 \\ \text{Rail 31 kg/m} \end{array}\right\} = 15.0 \text{ kN}$$

Crane data

Capacity 300 kN

Span 18.0 m

Weight of crane 280 kN

Weight of crab 70 kN

End carriage wheel centres 3.5 m

End clearance 300 mm

Headroom 2.5 m

Minimum hook approach 1.0 m

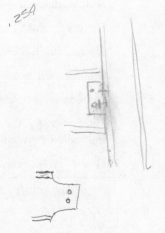

Wind loading. This is to be in accordance with CP3, Chapter V, Part 2 [2]. The location of the building is in the North East of England on the outskirts of a city.

4.2.3 Roof truss design

(a) Truss loads and analysis
The purlins will be located at the node points of the truss with a nominal spacing of 1903 mm.

The truss is to be fabricated in one piece using structural hollow sections in Grade 43 Steel.

The truss is analysed by the force diagram method. The dead imposed and wind loads are considered separately.

Dead loads

Internal panel points = 0.37 x 1.903 x 6 = 4.22 kN.

The truss loading and force diagram are shown on Fig. 4.4. The member loads are tabulated in Table 4.2.

Imposed loads

Internal panel points = 0.75 x 9460 x 6 x 1.903/9516.7 = 8.51 kN.

The member loads are found by proportion from the dead load analysis. These loads are tabulated in Table 4.2.

Wind loads

Location – North East England.

Basic wind speed – $V = 45$ m/s.

Factors S_1 and S_3 are both 1.0.

The building is sited on the outskirts of a city with obstructions up to 10 m high.

Ground roughness – Category 3.

Building size – horizontal dimension greater than 50 m – Class C.

The factors S_2, the design wind speed and the dynamic pressures for the roof and walls are given in Table 4.1.

Table 4.1   Wind pressures on roof and walls

	Height (m)	Factor S_2	Design wind speed V_s (m/s)	Dynamic pressure, q (kN/m^2)
Roof	10.2	0.694	45 x 0.694 = 31.23	$0.613 V_s^2/10^3$ = 0.598
Walls	9.2	0.68	45 x 0.68 = 30.6	= 0.574

The external pressure coefficients, C_{pe}, for the roof and walls are shown in Fig. 4.2. These are taken from Table 8 of CP3, Chapter V, Part 2. The internal pressure coefficients are taken from Appendix E. Here, for the case where there is only a negligible probability of a dominant opening occurring during a severe storm, C_{pi}, is taken as the more onerous of the values +0.2 or −0.3. The wind suction normal to the roof slope = $q(C_{pe} - C_{pi})$ kN/m^2.

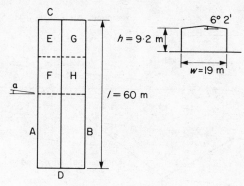

Figure 4.2

The pressure coefficients and wind loads for maximum uplift for the two cases where the wind angle is 0° and 90° are shown in Fig. 4.3(a) and (b), respectively. The panel point loads on the truss are:

(i) Wind angle α = 0°

  Windward slope = 0.694 x 6 x 1.903 = 7.92 kN

  Leeward slope = 0.359 x 6 x 1.903 = 4.1 kN

(ii) Wind angle α = 90°

  Both slopes = 0.598 x 6 x 1.903 = 6.83 kN

The truss loading and force diagrams for the two wind load cases are shown in Figs. 4.5 and 4.6. The wind loads are tabulated in Table 4.2.

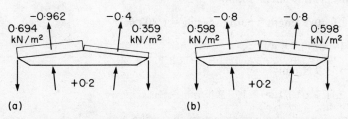

Figure 4.3 (a) Wind angle α = 0°. (b) Wind angle α = 90°.

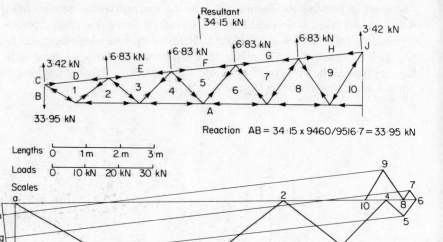

Reaction AB = 34·15 x 9460/9516·7 = 33·95 kN

Figure 4.4

(b) Design of the truss members
The design for the truss is made in Grade 43 steel.

Top chord members 5-F, 7-G, 9-H

Compression force = 231.3 kN

Tension force   = 46.1 kN

Try 100 x 100 x 5 RHS.

A = 18.9 cm^2

r = 3.87 cm

Purlins provide lateral support.

The effective length = purlin spacing = 1.903 m.

l/r = 1903/38.7 = 49.1

p_c = 133.9 N/mm^2

f_c = 231.3 x 10/18.9 = 122.3 N/mm^2. Satisfactory.

The section will also be satisfactory in tension.

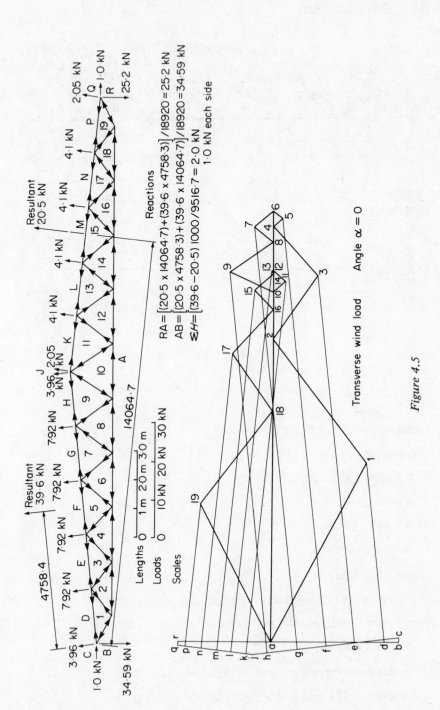

Resultant
39·6 kN

4758·4

3·96 kN

1·0 kN

7·92 kN   7·92 kN   7·92 kN   7·92 kN   3·96 2·05
                                            kN   kN

Resultant
20·5 kN

4·1 kN   4·1 kN   4·1 kN   4·1 kN   2·05 kN
                                        1·0 kN

25·2 kN

34·59 kN

14064·7

Lengths 0 1m 20m 30 m
Loads   0 10 kN 20 kN 30 kN
Scales

Reactions

$R_A = [(20\cdot5 \times 14064\cdot7) + (39\cdot6 \times 4758\cdot3)]/18920 = 25\cdot2$ kN

$A_B = [(20\cdot5 \times 4758\cdot3) + (39\cdot6 \times 14064\cdot7)]/18920 = 34\cdot59$ kN

$\Sigma H = [(39\cdot6 - 20\cdot5) \cdot 1000/9516\cdot7 = 2\cdot0$ kN

1·0 kN each side

Transverse wind load   Angle $\alpha = 0$

Figure 4.5

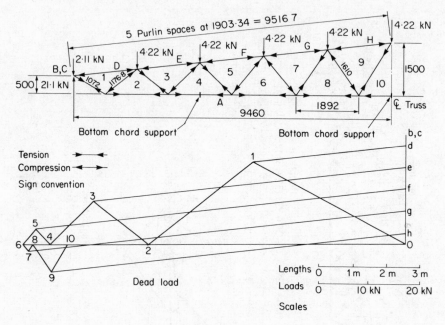

Figure 4.6

Top chord members 1-D, 3-E

Compression force = 189 kN

Try 100 x 100 x 4 RHS.

A = 15.3 cm^2

r = 3.91 cm

l/r = 1903/39.1 = 48.7

p_c = 134.3 N/mm^2

f_c = 189 x 10/15.3 = 123.5 N/mm^2. Satisfactory.

Bottom chord members 10-A, 8-A, 6-A, 4-A

Member 8-A tension force   = 224.3 kN

Member 6-A compression force =   43.9 kN

A required = 224.3 x 10/155   =   14.5 cm^2

Try 90 x 90 x 5 RHS.

Table 4.2   Summary of member forces (kN)

Member	Dead load	Imposed load	Wind angle 0°	Wind angle 90°	Dead + imposed	Dead + wind angle 0°	Dead + wind angle 90°
1-D	−30.5	− 61.5	46.0	46.5	− 92.0	15.5	16.0
3-E	−62.7	−126.4	96.5	99.6	−189.1	33.9	36.9
5-F	−76.7	−154.6	112.8	119.1	−231.3	36.1	42.4
7-G	−75.9	−153.0	111.0	112.0	−228.9	35.1	46.1
9-H	−71.0	−143.1	100	114.5	−214.1	29.0	43.5
11-K	−71.0	−143.1	97.4	110.8	−214.1	26.4	39.8
13-L	−75.9	−153.0	99.4	118.0	−228.9	23.5	42.1
15-M	−76.7	−154.6	94.3	115.3	−231.3	17.6	38.6
17-N	−62.7	−126.4	77.3	95.8	−189.1	14.6	33.1
19-P	−30.5	− 61.5	57.6	42.5	− 92.0	27.1	12.0
1-A	34.2	68.9	− 53.1	− 52.4	103.1	−18.9	−18.2
2-A	51.3	103.4	− 79.2	− 80.6	154.7	−27.9	−29.3
4-A	70.9	142.9	−108.2	−111.5	213.8	−37.3	−40.6
6-A	76.5	154.2	−112.3	−120.4	230.7	−35.8	−43.9
8-A	74.4	149.9	−104.3	−116.4	224.3	−29.9	−42.0
10-A	67.3	135.7	− 91.3	−104.6	203.0	−24.0	−37.3
12-A	74.4	149.9	− 96.8	−116.4	224.3	−22.4	−42.0
14-A	76.5	154.2	− 95.7	−120.4	230.7	−19.2	−43.9
16-A	70.9	142.9	− 86.7	−111.5	213.8	−15.8	−40.6
18-A	51.3	103.4	− 60.6	− 80.6	154.7	− 9.3	−29.3
19-A	34.2	68.9	− 41.3	− 52.4	103.1	− 7.1	−18.2
1-2	−26.3	− 53.0	41.2	43.7	− 79.3	14.9	17.4
2-3	13.7	27.6	− 20.4	− 21.7	41.3	− 6.7	− 8.0
3-4	−12.2	− 24.6	17.4	19.2	− 36.8	5.2	7.0
4-5	4.3	8.7	− 3.2	− 6.8	13.0	0.9	− 2.5
5-6	− 4.0	− 8.1	2.8	6.2	− 12.1	− 1.2	2.2
6-7	− 1.7	− 3.4	6.8	3.2	− 5.1	+ 5.1	1.5
7-8	1.6	3.2	− 6.5	− 3.4	4.8	− 4.9	− 1.8
8-9	− 6.5	− 13.1	13.9	10.5	− 19.6	7.4	4.0
9-10	6.3	12.7	− 13.0	− 10.5	19.0	− 6.7	− 4.2
10-11	6.3	12.7	− 4.4	− 10.5	19.0	1.9	− 4.2
11-12	− 6.5	− 13.1	4.2	10.5	− 19.6	− 2.3	4.0
12-13	1.6	3.2	0.9	− 3.4	4.8	2.5	− 1.8
13-14	− 1.7	− 3.4	− 0.6	3.2	− 5.1	− 2.3	1.5
14-15	− 4.0	− 8.1	6.5	6.2	− 12.1	2.5	2.2
15-16	4.3	8.7	− 7.0	− 6.8	13.0	− 2.7	− 2.5
16-17	−12.2	− 24.6	15.7	19.2	− 36.8	3.5	7.0
17-18	13.7	27.6	− 18.5	− 21.7	41.3	− 4.8	− 8.0
18-19	−26.3	− 53.0	30.8	43.7	− 79.3	4.5	17.4

Sign convention: + tension, − compression.

$A = 16.9 \ cm^2$

$r = 3.46 \ cm$

Location of the bottom chord supports are shown in Figs. 4.1 and 4.4. These are tied into the bracing in the top chord in the end bays.

Effective length = 3 x 1892 = 5676.

l/r = 5676/34.6   = 164

p_c = 34 + 25%   = 42.5 N/mm^2

f_c = 43.9 x 10/16.9 = 26.0 N/mm^2

Section is satisfactory in both tension and compression.

Bottom chord members 2-A, 1-A

Member 2-A   Tension force   = 154.7 kN

Compression force = 29.3 kN

A required = 154.7 x 10/155   = 10 cm^2

Try 90 x 90 x 3.6 RHS.

A = 12.4   cm^2

r =   3.52 cm

Effective length perpendicular to the plane of the truss (see Fig. 4.4), = 1072 + 1892 + 946 = 3910

l/r = 3910/35.2   = 110

p_c = 69 + 25%   = 86 N/mm^2

f_c = 29.3 x 10/12.4   = 23.6 N/mm^2

A smaller member could be used. However, this member will be selected to keep the bottom chord uniform in size.

Internal members

Member 1-2 Compression force = 79.3 kN

Try 50 x 50 x 4 RHS.

A = 7.28 cm^2

r = 1.87 cm

l/r = 0.85 x 1176.8/18.7   = 53.5

p_c = 130.5 N/mm^2

f_c = 79.3 x 10/7.28   = 109 N/mm^2. Satisfactory.

50 x 50 x 3.2 RHS   too light.

Members 8-9 Compression force = 19.6 kN

Try 50 x 50 x 3.2 RHS.

$A = 5.94 \text{ cm}^2$

$r = 1.91 \text{ cm}$

$l/r = 1610/19.1$ $\qquad = 84.1$

$p_c = 99 \text{ N/mm}^2$

$f_c = 19.6 \times 10/5.94$ $\qquad = 33.2 \text{ N/mm}^2$

A smaller member could be used.

All internal members except member 1-2 will be 50 x 50 x 3.2 RHS.

(c) Arrangement of the truss

The general arrangement of the truss with the member sections is shown in Fig. 4.7a. The truss support detail is also shown (Fig. 4.7b).

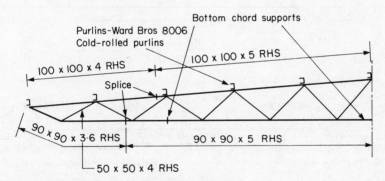

(a)

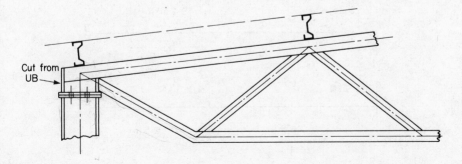

(b)

Figure 4.7   (a) Truss arrangement and member sizes. (b) Truss support detail.

4.2.4 Crane girder design

(a) Crane wheel loads, moments, shear
The crane wheel loads are shown in Fig. 4.8(a).

(i) Maximum load per wheel at A including impact of 25% = (280/2 + 370 x
= (280/2 + 370 x 17/18) 1.25/2   = 305.9 kN

Light side load per wheel at B = 650 x 1.25/2 − 305.9 = 100.35 kN.

Horizontal surge assumed, divided between four wheels

= 370 x 0.1/4   = 9.25 kN.

(ii) Maximum vertical bending moment. Place the loads on the beam as shown
in Fig. 4.8(b), i.e. with one wheel load and the centre of gravity of the loads
equidistant about the centre of the girder.

R_A = 7.5 + 305.9(5.625 + 2.125)/6   = 402.6 kN

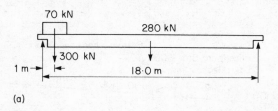

(a)

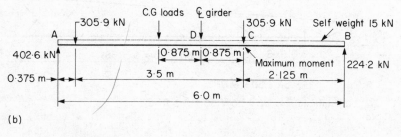

(b)

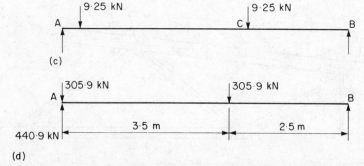

(c)

(d)

Figure 4.8   (a) Crane loads. (b) Vertical Loads − maximum moment. (c) Horizontal loads − maximum moment. (d) Loads causing maximum vertical shears.

$R_B = 15 + (2 \times 305.9) - 402.6$ = 224.2   kN

$M_C = (224.2 \times 2.125) - 15 \times 2.125^2/12$ = 470.8   kN m

Check case where one wheel load is placed at centre of girder.

$M_D = 305.9 \times 6/4 + 15 \times 6/8$ = 470.1   kN m

The case shown in the figure gives the maximum moment.

(iii) Maximum horizontal bending moment. See Fig. 4.8(c).

$M_C = 9.25(0.375 + 3.875)2.125/6$ =   13.92 kN m

(iv) Maximum shear. See Fig. 4.8(d).

$R_A = 305.9 + 305.9 \times 2.5/6 + 7.5$ = 440.9   kN

(b) Crane girder

The trial section for the crane girder is shown in Fig. 4.9(a). This is: 610 x
229 UB 113 + 305 x 89] 42.

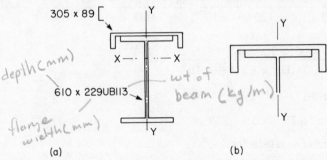

(a)   (b)

Figure 4.9   (a) Crane girder. (b) Surge girder.

The properties from the *Structural Steelwork Handbook* are:

$I_{XX} = 120699 \text{ cm}^4$,	$Z_B = 3158 \text{ cm}^3$
$r_{YY} = 7.2 \text{ cm}$,	$Z_{YY} = 568 \text{ cm}^3$
$Z_T = 5130 \text{ cm}^3$,	$D/T = 29.6$.

Allowable compressive stress:

$l/r_{YY} = 6000/72 \ = 83.3$

$p_{bc} = 165 + 10\% = 181.5 \text{ N/mm}^2$, Table 3(a) of BS449.

Vertical and horizontal stresses are to be added here:

Allowable average shear stress $p'_q = 100 \text{ N/mm}^2$.

Allowable tensile stress p_{bt} $= 165 \text{ N/mm}^2$

Compressive stress in top flange:

$$f_{bc} = \frac{470.8 \times 10^3}{5130} + \frac{13.92 \times 10^3}{568} = 91.8 + 24.5 = 116.3 \text{ N/mm}^2$$

Tensile stress in bottom flange:

$$f_{bt} = \frac{470.8 \times 10^3}{3158} \qquad = 149.1 \text{ N/mm}^2.$$

$$\text{Shear stress} = \frac{440.9 \times 1000}{607.3 \times 11.2} \qquad = 64.82 \text{ N/mm}^2.$$

The stresses are satisfactory.

Check the deflection at the centre of the girder. The formula for the deflection at the centre of the girder due to crane wheel loads is given in Section 3.4.6.

$$\delta = \frac{305.9 \times 10^3 \times 6000^3}{48 \times 2 \times 10^5 \times 120\,699 \times 10^4} \left[\frac{3(375 + 2125)}{6000} - \frac{4(375^3 + 2125^3)}{6000^3} \right]$$

$$+ \frac{5 \times 15\,000 \times 6000^3}{384 \times 2 \times 10^5 \times 120\,699 \times 10^4}$$

$$= 5.96 + 0.17 = 6.13 \text{ mm}$$

$$\delta/\text{span} = 6.13/6000 = 1/979$$

The deflection is satisfactory.

4.2.5 Crane column design

(a) Loading and analysis

Dead load − roof, column, crane girder and wall.

Stanchion cap	$= 0.37 \times 6 \times 9.516$	$= 21.12 \text{ kN}$
Crane girder level	$= 21.12 + 1.5 + (0.3 \times 6 \times 3.5)$	$= 28.92 \text{ kN}$
Stanchion base	$= 28.92 + 9 + (0.3 \times 6 \times 5.7) + 15$	$= 63.18 \text{ kN}$

The eccentric loads at crane girder level cause moments in the frame. The eccentricities for the assumed member sizes are shown in Fig. 4.10(b).

$$\text{Moment} = (28.92 \times 0.157) - (15 \times 0.298) \qquad = 0.07 \text{ kN m}$$

This is small and may be neglected. The column loads are shown Fig. 4.10(a).

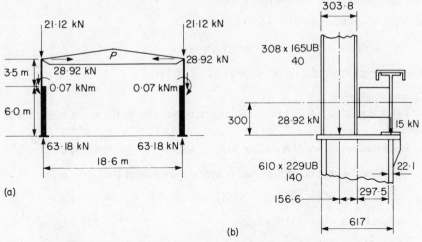

Figure 4.10 (a) Frame and loading. (b) Detail of crane girder.

Imposed load on roof

Load at stanchion cap = 0.75 x 9.46 x 6   = 42.57 kN.

Moment at crane girder level M = 42.56 x 0.157   = 6.68 kN m.

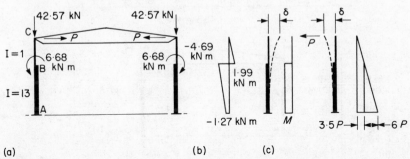

Figure 4.11 (a) Frame and loading. (b) Bending moment diagram. (c) Break-down of the column actions.

The frame, loading and analysis are shown in Fig. 4.11 where a force P is set up in the truss. The ratios of the inertias for the two portions of the column for the assumed sizes are shown in Fig. 4.11(a). Equate deflections at the column cap.

$$6M \times 6.5 \times 1/13 = 3.5P \times 1/2 \times 3.5 \times 2/3 \times 3.5 + 3.5P \times 6 \times 6.5 \times 1/13$$

$$+ 6P \times 1/2 \times 6 \times 7.5 \times 1/13$$

$$3M = 14.29P + 10.5P + 10.38P \qquad = 35.17P$$

$$P = 0.0853M = 0.0853 \times 6.68 \qquad = 0.57 \text{ kN}$$

$M_{BC} = 0.57 \times 3.5 \qquad = 1.99 \text{ kN m}$

$M_{BA} = -(6.68 - 1.99) = -4.69 \text{ kN m}$

$M_A = (9.5 \times 0.57) - 6.68 = -1.27 \text{ kN m}$

The bending moment diagram is shown in Fig. 4.11(b).

Crane wheel loads

Referring to Fig. 4.12(a) the column reactions due to the crane wheel loads are:

Maximum wheel loads $R_B = 2 \times 305.9 \times 4.25/6$ $\qquad = 433.4 \text{ kN}$

Check case where one wheel load is over the column at B

$R_B = 305.9(1 + 2.5/6)$ $\qquad = 433.4 \text{ kN}$

Moment at B, $M_B = 433.4 \times 0.298$ $\qquad = 129.2 \text{ kN m}$

Light side wheel loads $R_E = 100.35 \times 433.4/305.9$ $\qquad = 142.2 \text{ kN}$

Moment at E $\qquad M_E = 142.2 \times 0.298$ $\qquad = 42.4 \text{ kN m}$

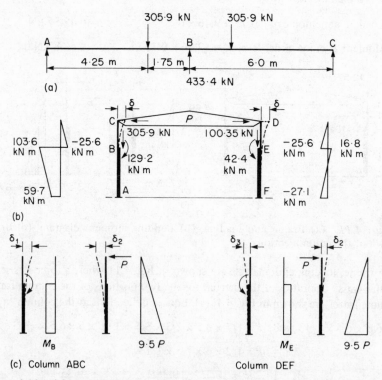

Figure 4.12 (a) Crane wheel loads. (b) Frame loading and column bending moment diagrams. (c) Breakdown of column actions.

The applied loads and moments are shown Fig. 4.12(b). The breakdown of the column actions is shown in Fig. 4.12(c). The results from the analysis for imposed loads are used to give:

$$\delta = \delta_1 - \delta_2 = 3M_B - 35.17P = \delta_2 - \delta_3 = 35.17P - 3M_E$$

$$P = 3(129.2 + 42.4)/(2 \times 35.17) \qquad\qquad = 7.32 \text{ kN}.$$

Column ABC	$M_{BC} = -7.32 \times 3.5$	$= -25.6$ kN m
	$M_{BA} = 129.2 - 25.6$	$= 103.6$ kN m
	$M_A = -(7.32 \times 9.5) + 129.2$	$= 59.7$ kN m.
Column DEF	$M_{ED} =$	$= -25.6$ kN m
	$M_{EF} = 42.4 - 25.6$	$= 16.8$ kN m
	$M_F = -(7.32 \times 9.5) + 42.4$	$= -27.1$ kN m.

The bending moment diagrams for the columns are shown in Fig. 4.12(b).

Crane surge loads

The column reactions due to the crane surge may be found in the same way as the reaction for the crane wheel loads. See Fig. 4.12(a).

$R = 9.25 \times 2 \times 4.25/6$	$= 13.1$ kN
$M_B = 13.1 \times 0.3$	$= 3.93$ kN m
$M_A = 13.1 \times 6.3$	$= 82.53$ kNm

The frame loads and moments are shown in Fig. 4.13.

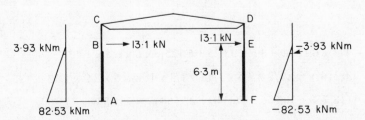

Figure 4.13   Frame loading and column bending moment diagrams.

Wind loads

Refer to the truss design above. The wind loads for the two cases of internal pressure and internal suction for wind angle 0° are shown in Fig. 4.14.

The analysis for the wind load on the walls is made for the case of internal pressure. See Fig. 4.15; consider the column subjected to w per unit length and a load P at top. The deflections at the top using the moment area theorem are proportional

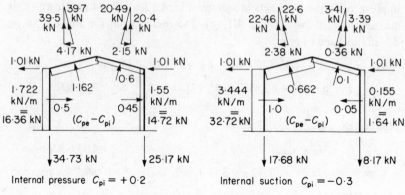

Internal pressure $C_{pi} = +0.2$ Internal suction $C_{pi} = -0.3$

Figure 4.14   Wind loads.

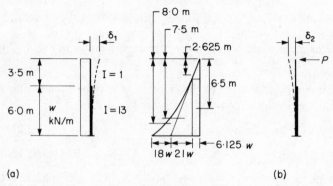

(a)   (b)

Figure 4.15   Deflections at the top of the column. (a) Uniformly distributed load. (b) Point load at top.

to:

$$\delta_1 = 1/3 \times 6.125w \times 3.5 \times 2.625 + 6.125w \times 6 \times 6.5 \times 1/13$$

$$+ 21w \times 6 \times 1/2 \times 7.5 \times 1/13 + 1/3 \times 18w \times 6 \times 8.0 \times 1/13$$

$$= (18.758 + 18.375 + 36.346 + 22.154)w = 95.633w$$

$\delta_2 = 35.17P$ (refer to imposed loads above)

For the building frame, the loads are shown in Fig. 4.16 for the case of internal wind pressure. The load P is given by:

$$1.72 \times 95.63 - 35.17P = 1.55 \times 95.633 + 35.17P$$

$$P = 0.234 \text{ kN}.$$

Column ABC $M_B = 1.722 \times 3.5^2/2 - (1.01 + 0.234)3.5$ $= 6.19$ kN m

$M_A = 1.722 \times 9.5^2/2 - (1.0 + 0.234)9.5$ $= 65.89$ kN m.

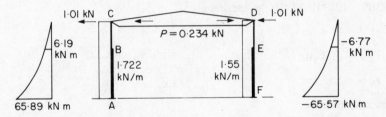

Figure 4.16 Frame loading and column bending moment diagrams. Wind – internal pressure case.

Column DEF $M_E = 1.55 \times 3.5^2/2 - (1.01 - 0.234)3.5 = -6.77$ kN m

$\qquad\qquad M_F = 1.55 \times 9.5^2/2 - (1.01 - 0.234)9.5 = -62.57$ kN m.

The bending moment diagrams for the columns are also shown in the figure. For the internal suction case, the frame loads are shown in Fig. 4.17. Here the load P is given by:

$3.444 \times 95.633 - 35.17P = 35.17P - 0.155 \times 95.633$

$P = 4.893$ kN.

Column ABC $M_B = 3.44 \times 3.5^2/2 - (1.01 + 4.893)3.5 = 0.43$ kN m

$\qquad\qquad M_A = 3.444 \times 9.5^2/2 - (1.01 + 4.893)9.5 = 99.33$ kN m.

Column DEF $M_E = 0.155 \times 3.5^2/2 - (4.893 - 1.01)3.5 = -12.64$ kN m

$\qquad\qquad M_F = 0.155 \times 9.5^2/2 - (4.893 - 1.01)9.5 = -29.9$ kN m.

The bending moment diagrams are also shown in Fig. 4.17.

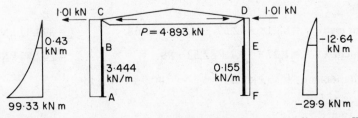

Figure 4.17 Frame loading and column bending moment diagrams. Wind – internal suction case.

Design of roof leg The design data are:

Axial load	$= 28.92 + 42.57 - 8.17$	$= 63.32$ kN
Moment	$= -1.99 - 25.6 - 3.94 - 12.64$	$= -44.17$ kN
Wind moment $= 12.64$ kN m		$= 28\%$ of the total.

Try 305 x 127 UB 37. The properties are:

$$A = 47.4 \text{ cm}^2, \qquad\qquad Z_{XX} = 470.3 \text{ cm}^3$$

$$r_{XX} = 12.3 \text{ cm}, \qquad\qquad D/T = 28.4$$

$$r_{YY} = 2.58 \text{ cm}, \qquad\qquad I_{XX} = 7143 \text{ cm}^3.$$

Allowable stresses:

$$l/r_{XX} = 1.5 \times 3500/123 \qquad\qquad = 42.7$$

$$l/r_{YY} = 3500/25.8 \qquad\qquad\qquad = 135.7$$

$$p_c = 48.3 \text{ N/mm}^2$$

$$p_{bc} = 120 \text{ N/mm}^2.$$

Actual stresses:

$$f_c = 63.32 \times 10/47.4 \qquad\qquad = 13.4 \text{ N/mm}^2$$

$$f_{bc} = 44.17 \times 10^3/470.3 \qquad\qquad = 93.9 \text{ N/mm}^2.$$

Combined:

$$\frac{f_c}{p_c} + \frac{f_{bc}}{p_{bc}} = \frac{13.8}{48.3} + \frac{93.9}{120} \qquad\qquad = 1.05 < 1.25.$$

The section is satisfactory.

Design of crane portion

The design data for the load case dead + imposed + crane wheel loads + surge + wind internal suction are:

(i) *base*

Axial load	= 63.18 + 42.57 + 305.9 − 17.68	= 393.97 kN
Moment	= 1.27 + 59.7 + 82.53 + 99.33	= 242.83 kN m
Wind moment = 99.33 kNm	= 40.9% of the total.	

(ii) *crane girder*

Axial load	= 28.92 + 15 + 42.57 + 305.9	= 392.39 kN
Moment	= −4.69 + 103.6 + 3.95	= 102.9 kN m.

Try 610 x 229 UB 101.

$$A = 129 \text{ cm}^2, \qquad\qquad Z_{XX} = 2509 \text{ cm}^3$$

$$r_{XX} = 24.2 \text{ cm}, \qquad\qquad D/T = 40.7$$

$$r_{YY} = 4.54 \text{ cm}, \qquad\qquad I_{XX} = 75\,549 \text{ cm}^4.$$

Allowable stresses:

$l/r_{XX} = 1.5 \times 6000/242$ $= 37.2$

$l/r_{YY} = 0.85 \times 6000/45.4$ $= 112.3$

$\quad p_c = 66.7 \text{ N/mm}^2$

$\quad p_{bc} = 144 \text{ N/mm}^2.$

Actual stresses:

$\quad f_c = 393.97 \times 10/129$ $= 30.54 \text{ N/mm}^2$

$\quad f_{bc} = 242.83 \times 10^3/2509$ $= 96.78 \text{ N/mm}^2.$

Combined:

$$\frac{f_c}{p_c} + \frac{f_{bc}}{p_{bc}} = \frac{30.54}{66.7} + \frac{96.8}{144} \qquad = 1.14 < 1.25.$$

The section is satisfactory.

The ratio of inertias for crane and roof parts is:

75 549/7 143 = 10.61

This is to be compared with a value of 13 used in analysis. Check bearing stress under crane girder on flange of crane stanchion

$$f_b = \frac{(305.9 + 15)1000}{227.6 \times 14.8} = 95.3 \text{ N/mm}^2. \text{ Satisfactory.}$$

Column base plate

The design data for the various load cases are:

(i) dead + imposed + crane wheels + surge + wind internal suction. Heavy side column

$P = 393.97 \text{ kN}$

$M = 242.83 \text{ kN m}$

Wind moment 40.9%;

(ii) dead + crane wheels + surge + wind internal suction. Light side column

$P = 63.18 + 100.35 - 34.73$ $= 129.8 \text{ kN}$

$M = -27.1 - 82.25 - 62.57$ $= -171.92 \text{ kN m}$

Wind moment is 36.5% of the total moment;

(iii) dead + wind internal pressure

$P = 63.18 - 25.17$ $\qquad$ = 38.01 kN

$M = -62.57$ kN m.

The allowable stresses including an increase for wind are:
concrete, grade 25 — allowable bearing pressure = 4.7 + 25% = 5.87 N/mm²
bolts, grade 4.6 — allowable tensile stress = 120 + 25% = 150 N/mm²

Try a base 1000 mm x 550 mm. The arrangement is shown in Fig. 4.18.

$$x = \left[\frac{15 \times 5.87}{(15 \times 5.87) + 150} \right] 950 \qquad = 351.4 \text{ mm}$$

$z = 950 - 351.4/3$ $\qquad$ = 832.9 mm

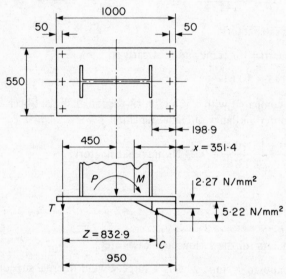

Figure 4.18

Case (i) loads
Take moments about the centre of the bolts in tension.

$M' = 242.8 + (393.9 \times 0.45)$ $\qquad$ = 419.8 kN m

$C = 419.8/0.833$ $\qquad$ = 504 kN

$f_c = 504 \times 2 \times 10^3/(351.4 \times 550)$ $\qquad$ = 5.22 N/mm² safe

$T = 504 - 393.9$ $\qquad$ = 110.1 kN

Case (ii) loads

$$M' = 171.92 + (129.8 \times 0.45) \qquad = 234.8 \text{ kN m}$$

$$C = 234.8/0.833 \qquad\qquad\qquad = 281.9 \text{ kN}$$

$$T = 281.9 - 129.8 \qquad\qquad\quad = 152.1 \text{ kN}$$

Case (iii) loads

$$M' = 62.57 + (38.0 \times 0.45) \qquad = 79.67 \text{ kN m}$$

$$C = 79.67/0.833 \qquad\qquad\qquad = 95.6 \text{ kN}$$

$$T = 95.6 - 38.01 \qquad\qquad\quad = 57.63 \text{ kN}$$

Using three bolts, the tensile stress area

$$= 152.1 \times 10^3/3 \times 150 \qquad\qquad = 338 \text{ mm}^2$$

Provide 3 no. 24 mm diameter bolts, $A_{net}= 353 \text{ mm}^2$

The thickness of the base plate is determined by designing it to cantilever from the column face. For 1 mm wide strip:

$$M = 5.22 \times 198.9^2/3 + 2.27 \times 198.9^2/6 \qquad\qquad = 8.38 \times 10^4 \text{ N mm}$$

The allowable stress $= 165 + 25\% \qquad = 206 \text{ N/mm}^2$ or 185 N/mm^2.

Thickness $= (8.38 \times 10^4 \times 6/206)^{1/2} \quad = 49.4 \text{ mm}$.

Use 50 mm thick base slab.

The weld force between column and base plate is

$$= \frac{171.92 \times 10^6}{227.6 \times 587.4} - \frac{129.8 \times 10}{2 \times 227.6} = 1285 - 286 \qquad = 999 \text{ N/mm}.$$

2 no. 8 mm fillet welds are required − strength 1280 N/mm.

Base plate 1000 mm x 550 mm x 50 mm thick

H.D. bolts 6 no. 24 mm diameter

Weld   Flanges 8 mm fillet weld each side

            Web 5 mm fillet weld each side

4.2.6 Gable column and rafter design
The arrangement for the gable framing is shown in Fig. 4.19. The design is made

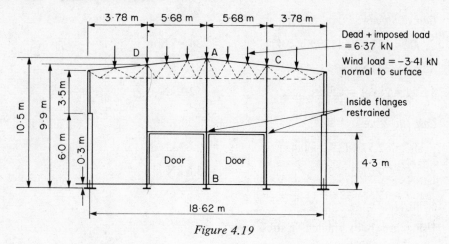

Figure 4.19

for the centre column **AB**. The axial load at mid height is:

Roof dead and imposed load = $(3 \times 6.37) + 1.4$ (beam)		= 20.51 kN
Roof wind load = $-(3 \times 3.41 \times 9460/9517)$		= −10.17 kN
Walls, insulation, rails = $0.3 \times 5.68 \times 5.25$		= 8.95 kN
Column, say 60 kg/m $\times$ 5.25 m		= 3.1 kN
Total		= 22.29 kN

Wind loads. The design is based on the wind load for the top of the wall. See Section 4.2.3, Fig. 4.2 where:

$C_{pe} - C_{pi} = 1.0$ or -0.8

Load	$= 0.574 \times 5.68 \times 10.2$	= 33.3 kN
Moment	$= 33.3 \times 10.5/8$	= 43.6 kN m.

Try 305 $\times$ 165 UB 40.

$Z_{XX} = 561.2 \text{ cm}^2$; $A = 51.5 \text{ cm}^2$, $r_{YY} = 3.85 \text{ cm}$, $D/T = 29.9$.

The inside flange is supported at the door top.

$l/r_{YY} = 6200/38.5 = 161$ (less than 180)

$p_c = 35 \text{ N/mm}^2$

$p_{bc} = 93 \text{ N/mm}^2$

$f_c = 22.39 \times 10/51.5$ $= 4.33 \text{ N/mm}^2$

$$f_{bc} = 43.6 \times 10^3/561.2 \qquad\qquad = 77.8 \text{ N/mm}^2$$

$$f_c/p_c + f_{bc}/p_{bc} = 4.33/35 + 77.8/93 \qquad\qquad = 0.96 < 1.25$$

All gable columns to be 305 x 165 UB 40.

Rafter AC: $M = 6.37 \times 5.68/3 \qquad\qquad = 12.1 \text{ kN m}$

$$Z = 12.1 \times 10^3/165 \qquad\qquad = 73.1 \text{ cm}^3.$$

Provide 254 x 102 UB 22 where $Z \qquad\qquad = 225.7 \text{ cm}^3.$

4.2.7 Bracing and sheeting rails
The roof and side wall bracing is shown on Fig. 4.20(a) and (b) respectively.

(a) Roof bracing
This is designed for the maximum loads on one gable. These are:

(i) wind on the gable end

Load at C   = 16.28 kN (see gable column design)

Loads at B, D   = 4.73 x 9.95 x 0.574/2 $\qquad\qquad$ = 13.51 kN

Loads at A, E   = 1.8 x 1.89 x 0.574 $\qquad\qquad$ = 1.95 kN;

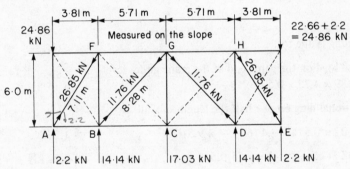

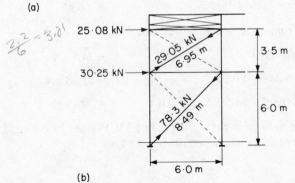

Figure 4.20   (a) Roof bracing and loading. (b) Side wall bracing and loading.

(ii) frictional drag on roof. See Clause 7.4 of CP3, Chapter V.

Part 2. Here $d = 60$ m, $h = 9.5$ m, $b = 18.92$ m

$$d/h = \frac{60}{9.5} = 6.32, d/b = \frac{60}{18.92} \qquad\qquad = 3.2.$$

$C_f' = 0.02$ for surfaces with corrugations across the wind direction.

For $h < b$, the load on one half of the load surface

$F' = 0.02 \times 0.590 \times 19.03(60 - 4 \times 9.5)/2$		= 2.5 kN.
Load at C	$= 2.5 \times 5.71/19.03$	= 0.75 kN
Loads at B, D	$= 2.5 \times 4.76/19.03$	= 0.63 kN
Loads at A, E	$= 2.5 \times 1.9/19.03$	= 0.25 kN

The loads are shown in Fig. 4.20(a). The maximum tension in member MF is 26.85 kN. Provide 50 x 50 x 6L. The safe load from the *Structural Steelwork Handbook* is 55 kN, where one 18 mm diameter hole has been deducted from the connected leg. All bracing members will be made the same size. The bracing is to be clipped to the purlins.

(b) Wall bracing

The loading is due to:

(i) wind load on the gable end at the crane girder level
$= 1.89 \times 4.75 \times 0.574$ = 5.15 kN;

(ii) frictional drag on one half the length of wall

$= 0.02 \times 0.574 \times 9.4 (60 - 4 \times 9.5)/2$	= 1.2 kN.
Load at eaves $= 1.2 \times 1.75/9.5$	= 0.22 kN
Load at crane girder $= 1.2 \times 4.75/9.5$	= 0.6 kN;

(iii) longitudinal surge from crane $- 5\%$ of static wheel loads

$= 0.05(140 + 370 \times 17/18)$	= 24.5 kN.
Load at eaves level $= 24.86 + 0.22$	= 25.08 kN
Load at crane girder level $= 24.5 + 5.15 + 0.6$	= 30.25 kN

The maximum tension force in the bracing is 78.3 kN. Provide 70 x 70 x 6L. Safe load allowing for a 22 mm diameter hole is 83 kN.

(c) Eaves strut

Load = 26.53 kN compression

Maximum l/r ratio = 250

Minimum $r = 6000/250 \times 10$ = 2.4 cm

Try 203 × 89 [× 29.78.

$A = 37.94 \text{ cm}^2$; r_{YY} = 2.64 cm

$l/r = 6000/26.4$ = 227.2

$p_c = 19 + 25\%$ = 23.7 N/mm^2

$f_c = 25.05 \times 10/29.78$ = 8.4 N/mm^2

Section is satisfactory.

(d) Sheeting rails

Wind load = 0.574 kN/m^2

Span 6.0 m, centres 1.78 m. See Fig. 4.21.

Provide Ward Bros. R 5562. Safe wind load = 0.83 kN/m^2.

The manufacturer provides the sheeting rail support system which is independent of the eaves strut.

4.2.8 Arrangement of steel framing
The details for the main frame and the framing plan members are shown in Fig. 4.21.

4.3 Simple design — two-storey building

4.3.1 Specification
A mailing company is to expand its office, storage and despatching space on a restricted site. The new structure will be erected over the existing premises which will be subsequently demolished. A section through the proposed structure is shown in Fig. 4.22. The floor providing office space is suspended from the roof truss.

Details of the building and construction are as follows:

frame spacing   = 4.5 m;

length of building = 36 m;

resistance to wind loading is to be provided by bracing in the roof, floor, side walls and gables;

imposed loading on the first floor which comprises office space with data processing equipment   = 3.5 kN/m^2;

roof — steel deck, insulation board, three layers of felt and chippings;

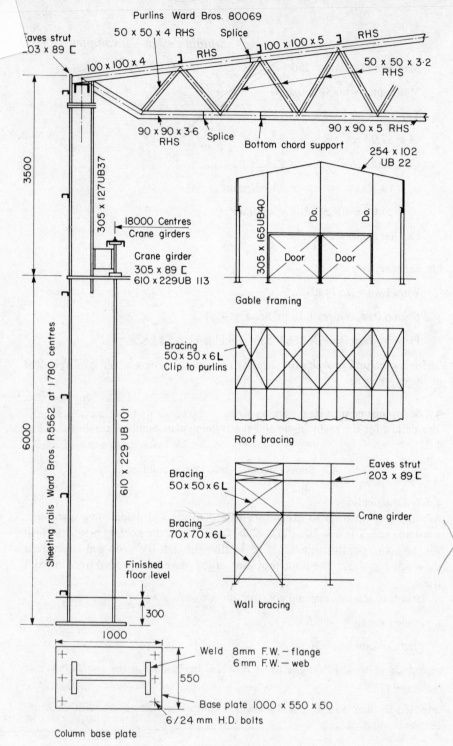

Figure 4.21

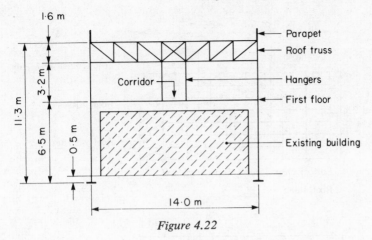

Figure 4.22

first floor – precast concrete floor units with screed and tiles.

Floor steel is protected by asbestos spray with suspended ceiling under. Columns and hangers are to be solid cased.

External walls are to be in 255 mm thick cavity brick construction with glazing along the sides.

The corridor walls in the office block are in 102 mm breeze block.

Other internal subdivision is to be in lightweight partitions. The end bays contain toilets, stairs and lifts. The walling here is in 225 mm brick. The detailed design in these bays is not part of the problem.

The following design work is to be carried out:

(a) draw complete framing plans for the structure;
(b) show suitable constructional details on sketches and estimate the loading for the roof, floor and walls;
(c) design the roof steel, floor steel, roof truss hangers and side column for an internal bay;
(d) show the main frame members and structural details on sketches.

Design calculations for the overall wind loading and bracing for the building are not part of the problem.

4.3.2 Framing plans
The framing plans showing all the main structural steel members are given in Fig. 4.23. Longitudinal wind girders in the roof and floor transmit lateral wind load to vertical bracing in the gables. Longitudinal wind is taken by bracing in the roof and side walls. No uplift will occur on the roof because of the high dead load, so lower chord bracing for the roof trusses is not required.

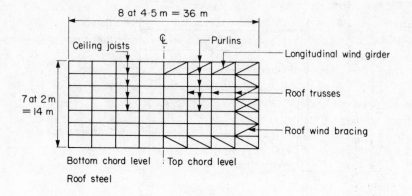

8 at 4·5 m = 36 m

Ceiling joists ⟶ Purlins ⟶ Longitudinal wind girder

7 at 2 m = 14 m

⟶ Roof trusses

⟶ Roof wind bracing

Bottom chord level · Top chord level

Roof steel

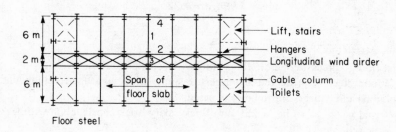

6 m

2 m

6 m

4
1
2
3

Span of floor slab

Lift, stairs
Hangers
Longitudinal wind girder
Gable column
Toilets

Floor steel

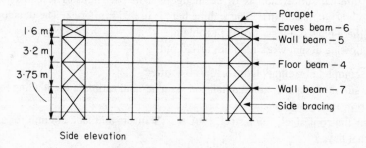

1·6 m

3·2 m

3·75 m

Parapet
Eaves beam — 6
Wall beam — 5

Floor beam — 4

Wall beam — 7
Side bracing

Side elevation

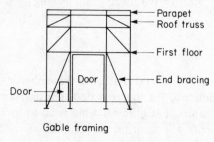

Parapet
Roof truss

First floor

Door

Door

End bracing

Gable framing

Figure 4.23

Joists attached to the lower chord of the roof truss are provided to support the first floor ceiling. A longitudinal beam is provided at mid height in the side wall to give the lower column length lateral support. This member also provides vertical support for the upper portion of the brick side walls and glazing and carries the wind load on the sides to the columns.

4.3.3 Constructional details and estimated loading

Constructional details for the roof, floors and external walls for an internal bay are shown in Fig. 4.24. The arrangement of the side wall is given showing the heights of brickwork and glazing. These details are used to estimate:

(a) weights per square metre for the roof, floor and external wall;
(b) weight per metre for the column, hanger, edge beam and wall beam.

These estimated weights are used in the design of the various members.

(a) Roof loading

	(kN/m^2)
Felt — three layers with chippings	0.4
Insulation board	0.22
Steel decking — 2.0 m span	0.28
Purlins	0.15
Truss — estimated weight 40 kN	0.74
Ceiling and joists	0.5
Services	0.1
Total dead load	2.39
Imposed load — flat roof with access	1.5
Total load	3.89

(b) Floor loading

The precast floor slab has to carry about 6 kN/m^2 over a span of 4.5 m. Use a Bison floor slab of 185 mm depth which weighs 2.44 kN/m^2 and can carry a safe load of 6.4 kN/m^2.

	(kN/m^2)
Rubber tiles	0.14
Screed — 30 mm	0.7
Precast slab	2.44
Steel and protection	0.3

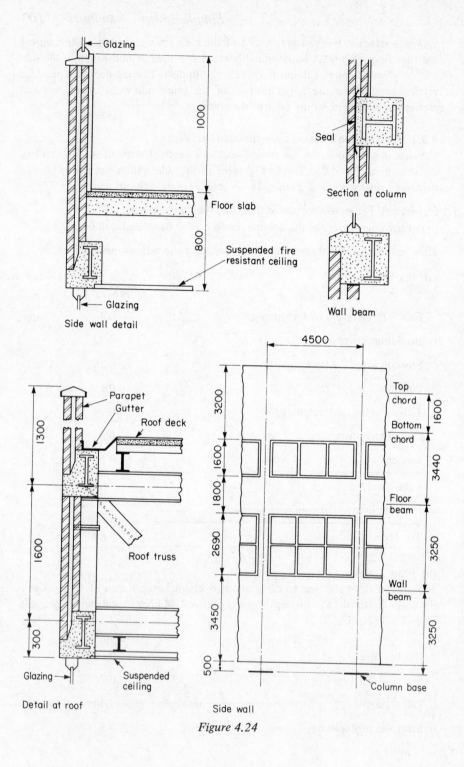

Glazing

1000

Floor slab

800

Suspended fire
resistant ceiling

Glazing

Side wall detail

Seal

Section at column

Wall beam

Parapet
Gutter

Roof deck

1300

1600

Roof truss

300

Glazing

Suspended
ceiling

Detail at roof

4500

Top
chord

Bottom
chord

1600

3200

1600

3440

1800

Floor
beam

2690

3250

Wall
beam

3450

3250

500

Column base

Side wall

Figure 4.24

Ceiling	0.3
Services	0.1
Lightweight partitions	1.0
Total dead load	4.98
Imposed load	3.5
Total load	8.48

(c) Corridor walls on the office floor

102 mm breeze block plastered both sides	2.4 kN/m^2

(d) External walls

Brickwork — 254 mm cavity wall plastered one side	5.1 kN/m^2
Glazing	0.3 kN/m^2

(e) Cased edge beam and wall beam

Steel beam 305 x 165 UB 40	0.4 kN/m
Casing 405 x 265	2.53 kN/m
Total	2.93 kN/m

(f) Cased column

Steel column 203 x 203 UC 60	0.6 kN/m
Casing 310 x 310	2.27 kN/m
Total	2.87 kN/m

(g) Cased hanger

Steel section 203 x 203 UC 40	0.4 kN/m
Casing 305 x 305	2.19 kN/m
Total	2.59 kN/m

4.3.4 Steelwork design

(a) Purlins and beams

The design for the purlins, floor beams and wall beams is given on the calculation sheets.

Member or mark	Loading and design calculations	Section and properties
Purlins	 Chippings, felt, insulation, deck, purlins and imposed load $W = (0.5 + 0.21 + 0.12 + 0.27 + 0.15 + 1.5)4.5 \times 2 = 24.8$ kN Imposed load $= 1.5 \times 4.5 \times 2 = 13.5$ kN $M = 24.8 \times 4.5/8 = 13.92$ kN m $Z = 13.92 \times 10^3/165 = 84.4$ cm^3 Deflection need not be checked	152 x 89 Joist 17.09 $Z = 115.6$ cm^3
Ceiling joist	 Ceiling, joists, services $W = 0.5 \times 4.5 \times 2 = 4.5$ kN Weight of a man 0.9 kN at mid span $M = 4.5 \times 4.5/8 + 0.9 \times 4.5/4 = 3.55$ kN m $f_b = 3.55 \times 10^3/74.94 = 47.4$ N/mm^3 $\delta = \dfrac{5 \times 4500^4}{384 \times 2.1 \times 475.9 \times 10^9} +$ $\dfrac{900 \times 4500^3}{48 \times 2.1 \times 475.9 \times 10^9} = 7.4$ mm $\delta/\text{span} = 7.4/4500 = 1/608$ $l/r_{YY} = 4500/17.2 = 262$ $p_b = 82$ N/mm^2 Table 3a of BS449 Section is satisfactory	Try 127 x 76 Joist 13.36 $Z = 74.94$ cm^3 $I = 475.9$ cm^4 $r_{YY} = 1.72$ cm $D/T = 16.7$ 127 x 76 Joist 13.36
Floor beam Mk 1	 $W = 8.48 \times 6 \times 4.5 = 229$ kN Imposed load $= 3.5 \times 6 \times 4.5 = 94.5$ kN $M = 229 \times 6/8 = 171.7$ kN m $Z = 171.7 \times 10^3/165 = 1040.6$ cm^3 Moment of intertia to limit deflection to span/360 $I = 2.34 \times 229 \times 62 = 19\ 291$ cm^4	457 x 152 UB 60 $Z = 1120$ cm^3 $I = 25464$ cm^4

Member or mark	Loading and design calculations	Section and properties
Floor beam Mk 2	Beam + 200 mm width of slab + corridor wall $W = (0.3 + 0.2 \times 2.44 + 2.4 \times 2.9)4.5$ $\qquad\qquad\qquad\qquad = 34.9$ kN $M = 34.9 \times 4.5/8 \qquad = 19.6$ kN m $Z = 19.6 \times 10^3/165 \qquad = 118.9$ cm^3 $I = 2.34 \times 34.9 \times 4.5 \qquad = 1654$ cm^4	254 × 102 UB 22 $Z = 225.4$ cm^3 $I = 2863$ cm^4
Floor beam Mk 3	$W = 7.48 \times 4.5 \times 2 \qquad = 67.3$ kN Imposed load $= 3.5 \times 4.5 \times 2 = 31.5$ kN $M = 67.3 \times 2/8 \qquad = 16.83$ kN m $Z = 16.83 \times 10^3/165 \qquad = 102$ cm^3 $I = 2.34 \times 67.3 \times 4 \qquad = 629.9$ cm^4	152 × 89 Joist 17.09 $Z = 115.6$ cm^3 $I = 881.1$ cm^4
Wall beam Mk 4	Glazing, cavity brickwall, beam casing $W = [(0.3 \times 1.6) + (5.1 \times 1.8)$ $\qquad + 2.93]4.5 \qquad = 56.7$ kN $M = 56.7 \times 4.5/8 \qquad = 31.9$ kN m Cased section $r_{YY} = 0.2(101.9 + 100)$ $\qquad\qquad\qquad\qquad = 40.38$ $\qquad l/r_{YY} = 4500/40.38$ $\qquad\qquad\qquad = 111.4$ $\qquad p_b = 146$ N/mm^2 Uncased section $\qquad l/r_{YY} = 4500/20.5$ $\qquad\qquad\qquad = 219.5$ $\qquad p_{b_3} = 62 + 50\% = 93$ N/mm^2 $f_b = 31.9 \times 10.3^3/350.7 \qquad = 91$ N/mm^2 $I = 2.34 \times 56.7 \times 4.5^2 \qquad = 2\,687$ cm^4 Section is satisfactory	Try 305 × 102 UB 28 $Z = 35.07$ cm^3 $I = 5\,415$ cm^4 $r_{YY} = 2.05$ cm $D/T = 34.7$ 305 × 102 UB 28
Wall beam Mk5	Cavity brick wall, beam, casing $W = [(5.1 \times 1.9) + 2.93]4.5 = 56.8$ kN Make eaves beam, Mk 6 and wall beam same. Loads on these beams are Mark 6 $W = [(5.1 \times 1.3) +$ $\qquad 2.93]4.5 \qquad = 43$ kN Mark 7 $W = [(0.3 \times 3.25) +$ $\qquad 2.93]4.5 \qquad = 17.6$ kN	305 × 102 UB 28

(b) Hanger

The loading on the hanger from the floor beams is shown in Fig. 4.25(a). The hanger is checked at the connection to the floor beams where the eccentric floor beam loads cause a moment in the member.

Total tension = 67.2 + 47.3 + (2 x 17.5) + 17.9 + 15.8   = 183.2 kN

Moment M_{YY} = (67.2 + 47.3 − 17.9 −15.8)0.104   = 8.4 kN m

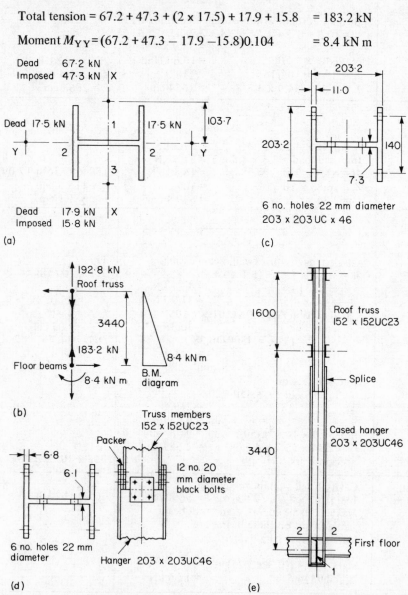

Figure 4.25 (a) Loading from floor beams. (b) Loads and moments in hanger. (c) Net section at floor beam connection. (d) Splice details. (e) Elevation of hanger.

The bending moment diagram for the hanger is shown in Fig. 4.25(b).

Tension at top of hanger = 183.2 + (2.59 x 3.7)   = 192.8 kN.

Try 203 x 203 UC 46. The net section allowing for 6 no. 22 mm diameter holes is shown in Fig. 4.25(c).

Net area $A = 58.9 - (4 \times 2.2 \times 1.1) - (2 \times 2.2 \times 0.73)$ = 45.91 cm^2

Net moment of inertia about YY axis:

$$I_{YY} = 1539 - (4 \times 2.2 \times 1.1 \times 7.0^2) \qquad = 1065 \text{ cm}^4.$$

Stresses on net section:

$f_t = 183.2 \times 10/45.91$ = 39.9 N/mm^2

$$f_{bt} = \frac{8.4 \times 10^6 \times 203.2}{2 \times 1065 \times 10^4} \qquad = 80.1 \text{ N/mm}^2.$$

Combined:

$$\frac{f_t}{p_t} + \frac{f_{bt}}{p_{bt}} = \frac{39.9}{155} + \frac{80.1}{165} \qquad = 0.745.$$

This is satisfactory. The lighter section 152 x 152 UC 37 is overstressed.

Bolts at splice.   Use 20 mm diameter black bolts with a single shear value of 25.1 kN per bolt.

Numbers of bolts = 183.2/25, say eight bolts.

Connection to truss. Try 152 x 152 UC 23 for truss member.
The net section is shown in Fig. 4.25(d)

The net area = $29.8 - (4 \times 2.2 \times 0.68) - (2 \times 2.2 \times 0.61)$ = 21.13 cm^2

Stress $f_t = 192.8 \times 10/21.13$ = 91.2 N/mm^2.

The proposed member is satisfactory.

The splice details are shown in Fig. 4.25(d). The arrangement of the hanger is shown in Fig. 4.25(e).

(c) Roof truss
The truss supports the roof and first floor. The reduction in imposed load allowed in design is 10%. See CP3, Chapter V, Part 1, Table 2 [6].

Hanger load = 67.2 + (17.5 x 2) + 17.9 + (3.7 x 2.5) + 0.9 (47.3 + 15.8) = 186.5 kN

Roof truss panel point loads:

top chord = (1.57 + 0.9 x 1.5)2 x 4.5   = 26.28 kN.

bottom chord = 0.82 x 2 x 4.5   = 7.38 kN.

114   Steel structures

The loading on the truss is shown in Fig. 4.26(a). The truss is analysed by joint resolution and the member forces are shown on the truss.

The design of the truss members is given below:

Diagonals BC, DE.   Force = 460.15 kN tension

$A = 460.15 \times 10/155 = 29.7 \text{ cm}^2$

This member is connected to the top chord by welding both flanges.

Use 152 x 152 UC 23   where $A = 29.8 \text{ cm}^2$.

Diagonal FG.   Force = 352.39 kN tension.

$A = 352.39 \times 10/155 = 22.7 \text{ cm}^2$

Use 152 x 76[17.88   where $A = 22.77 \text{ cm}^2$.

Vertical DC.   Force = 280.1 kN compression

Try 152 x 76[17.88.

$A = 22.77 \text{ cm}^2$

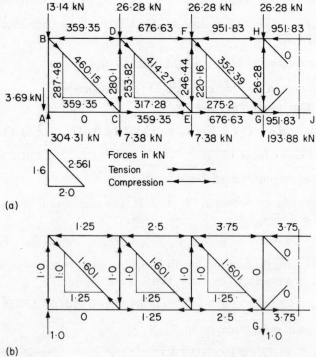

(a)

(b)

Figure 4.26   (a) Loads on truss and member forces. (b) Unit load applied at G — forces in members.

$r_{YY} = 2.24$ cm

$l/r_{YY} = 0.85 \times 1600/22.4$ $\qquad = 60.7$

$\quad p_c = 125.2$ N/mm^2

$\quad f_c = 280.1 \times 10/22.77$ $\qquad = 123$ N/mm^2

Use 152 x 76[17.88.

Vertical FE.   Force = 246.44 kN compression.

Use 152 x 76[17.88.

Vertical HG.   Force = 26.28 kN compression.

Use 152 x 152 UC 23.

This member is extended to splice on to hanger as shown in Fig. 4.25.

Bottom chord GJ.   Force = 951.83 kN tension.

$A = 951.83 \times 10/155$ $\qquad = 61.41$ cm^2

Use 2 no. 229 x 76] [36.06.

$A = 2 \times 33.2$ $\qquad = 66.4$ cm^2

Make members CE, EG the same size.

Bottom chord CE. Use 2 no. 127 x 64] [14.90.

Top chord HK.   Force = 951.83 kN compression.

Try 2 no. 229 x 76] [26.06 space at 152.4 mm back to back. The member is shown in Fig. 4.27.

$\quad A = 2 \times 33.2$ $\qquad = 66.4$ cm^2

$r_{XX} = 8.87$ cm

$I_{YY} = 2 \times 158.7 + 66.4 \times 9.62^2$ $\qquad = 6462.3$ cm^4

$r_{YY} = (6462.3/66.4)^{1/2}$ $\qquad = 9.86$ m

$l/r_{XX} = 0.7 \times 2000/8.87$ $\qquad = 15.8$

$l/r_{YY} = 2000/98.6$ $\qquad = 20.3$

$\quad p_c = 146.7$ N/mm^2

$\quad f_c = 951.83 \times 10/66.4$ $\qquad = 143.3$ N/mm^2.

Make the top chord uniform throughout.

Deflection due to total load
Check the deflection of the truss at the hanger connection. The strain energy

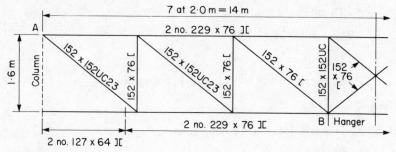

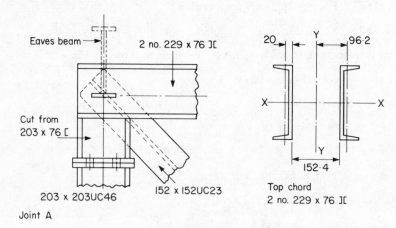

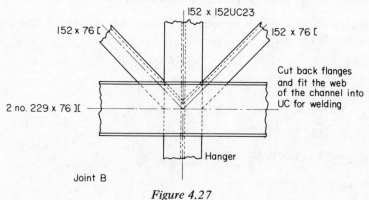

Figure 4.27

method is used in the calculation. The forces in the truss due to a unit load at
G are given in Fig. 4.26(b). The deflection calculation is given in Table 4.3.

$$\delta = 2 \times 5\ 661\ 584.7/2.1 \times 10^5 \quad = 53.9 \text{ mm}$$

$$\delta/\text{span} = 53.9/14\ 000 \quad\quad\quad = 1/259$$

Table 4.3 Deflection of truss

Member	Force, P (kN)	Force from unit load, U	Length, L (mm)	Area, A (mm²)	$\dfrac{PUL}{A}$
BD	−359.35	−1.25	2 000	6 640	135 297.4
DF	−676.63	−2.5	2 000	6 640	509 510.5
FH	−951.83	−3.75	2 000	6 640	1 075 109.1
HK	−951.83	−3.75	1 000	6 640	537 554.6
CE	359.35	1.25	2 000	6 640	135 297.4
EG	676.63	2.5	2 000	6 640	509 510.5
GJ	951.83	3.75	1 000	6 640	537 554.6
BC	460.15	1.601	2 561	2 980	633 117.1
DE	414.27	1.601	2 561	2 980	569 991.2
FG	352.39	1.601	2 561	2 277	634 543.6
CD	−280.1	−1.0	1 600	2 277	196 820.4
EF	−246.44	−1.0	1 600	2 277	173 168.2
GH	− 26.28	−1.0	1 600	2 980	14 110.1
					Σ5 661 584.7

The truss can be cambered to offset the dead load deflection. The calculations for this are not given. The camber should be about 40 mm.

Structural details for truss
The member sizes and the main details showing construction of the truss are given in Fig. 4.27.

(d) Column
The loads coming into the column from roof truss floor beams and wall beams are shown in Fig. 4.28. The reduction in imposed load allowed is 10%.

The stresses are checked at the following locations:

(i) *below floor level.* See Fig. 4.28(d).

Axial load $= (2 \times 21.5) + 304.3 + 0.6 + (4 \times 28.4) + 109.8 + 4.9 + 9.9$
$= 586.1$ kN

Moment $M_{XX} = 109.8 \times 0.202 \left[\dfrac{1/6.5}{1/65 + 1/3.44} \right]$ $= 7.67$ kN m

Trial section 203 x 203 UC 46.

Cased section with 50 mm cover − 303.2 mm x 303.2 mm.

For steel section:

$A = 58.8$ cm², $\quad r_{XX} = 8.81$ cm, $\quad D/T = 18.5$

$Z_{XX} = 449.2$ cm³, $\quad r_{YY} = 5.11$ cm,

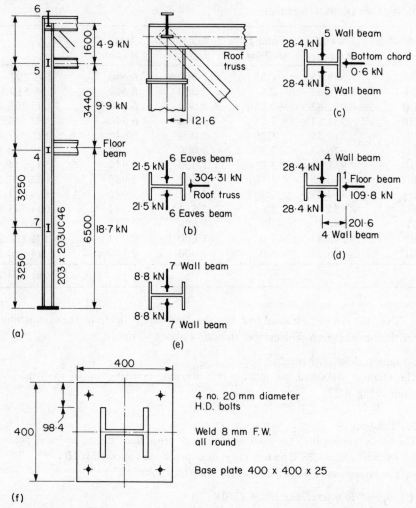

Figure 4.28 (a) Elevation of column. (b) Joint – roof truss to column. (c) Wall beam and bottom chord of truss. (d) Floor beam and wall beam connection. (e) Wall beam connection. (f) Column base plate.

$l/r_{XX} = 0.85 \times 6500/88.1$ = 62.7

$l/r_{YY} = 3250/51.1$ = 63.6

$p_c = 122 \text{ N/mm}^2$

Safe axial load on steel section = 58.8 × 122/10 = 717.4 kN.

For the cased section:

$r_{YY} = 0.2 \times 303.2$	$= 60.6$ mm
$l/r_{YY} = 3250/60.6$	$= 53.6$
l/r_{XX}	$= 62.7$
$p_c = 123$ N/mm^2	
$p_{bc} = 165$ N/mm^2	

$$\text{Safe load} = \frac{123 \times 58.8}{10} + \frac{303.2^2 \times 123}{0.19 \times 165 \times 10^3}$$

$$= 723.2 + 360.7 \qquad\qquad = 1083.9 \text{ kN.}$$

This is not greater than twice the safe load in the steel section.

Bending stress $f_{bc} = 7.67 \times 10^3/449.2$ $\qquad = 17.1$ N/mm^2.

Combined stresses on the cased section:

$$\frac{586.1}{1083.9} + \frac{17.1}{165} = 0.54 + 0.11 \qquad\qquad = 0.65.$$

The section is satisfactory.

(ii) Top chord level of the roof truss at the connection to the column.

Assume the roof truss reaction is applied at the edge of the plate. See Fig. 4.28(b). This is not designed as a cased section.

Axial load $= 304.3 + (21.5 \times 2)$	$= 347.3$ kN
Moment $= 304.3 \times 0.122$	$= 37.12$ kN m
$l/r_{YY} = 1600/51.1$	$= 31.3$
$p_c = 142$ N/mm^2	
$f_c = 347.3 \times 10/58.8$	$= 59.1$ N/mm^2
$f_{bc} = 37.12 \times 10^3/449.2$	$= 82.7$ N/mm^2

Combined
$$\frac{59.1}{142} + \frac{82.7}{165} \qquad\qquad = 0.917$$

The column section is satisfactory.

(e) Column base plate

Column base load $= 586.1 + 17.6 + 18.7$	$= 622.4$ kN.
For Grade 21 concrete, the safe bearing pressure	$= 4.2$ N/mm^2.

Using a square base, the side length

$$= (622.4 \times 10^3/4.2)^{1/2} \qquad\qquad = 384.9 \text{ mm.}$$

Make the base plate 400 mm x 400 mm.

Bearing pressure $= 622.4 \times 10^3/400^2 \qquad\qquad = 3.89 \text{ N/mm}^2.$

The projections beyond the column are each

$$= (400 - 203.2)/2 \qquad\qquad = 98.4.$$

The thickness of base plate is given by

$$t = \left[\frac{3 \times 3.89}{185} \left(98.4^2 - \frac{98.4^2}{4} \right) \right]^{1/2}$$

$$= 21.4 \text{ mm.}$$

Provide a base plate 400 mm x 400 mm x 25 mm thick and 4 no. 20 mm H.D. bolts.

Weld column to base plate — 6 mm fillet weld all round.

The column base is shown in Fig. 4.28 (f).

4.4 Coursework exercises

1. The cross-section through a factory building is shown in Fig. 4.29(a). The frames are at 5 m centres and the length of the building is 40 m. The purlin spacing to be used is shown in Fig. 4.29(b). The building loads are:

Roof — dead load measured on the slope	(kN/m^2)
sheeting 20 gauge on purlins at 1.76 m centres	0.1
insulation board	0.15
purlins 120 x 120 x 8L to BS449 Clause 45	0.1
roof truss (estimate)	0.1
Total dead load	0.45

Roof — imposed load measured on plan	0.75 kN/m²
Walls — sheeting, insulation board, sheeting rails	0.35 kN/m²
column	2.5 kN

Wind loading — CP3, Chapter V, Part 2 [2]. The location of the building is in the North East of England on the outskirts of a city.

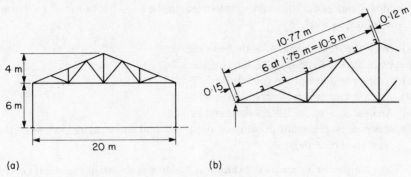

(a)   (b)

Figure 4.29   (a) Section. (b) Purlin spacing.

Carry out the following work:

(a) estimate the loading on an internal frame due to dead, imposed and wind loads;
(b) analyse and design the roof truss and columns using Grade 43 Steel;
(c) show the main structural details on a drawing
(d) calculate the weight of an internal frame and compare this with the weight of the rigid portal designed in Chapter 5.
2. The cross-section through an industrial building is shown in Fig. 4.30(a). The frames are at 6.5 m centres and the length of the building is 52 m. The crane wheel loads are shown in Fig. 4.30(b). Select suitable steel sheeting purlins

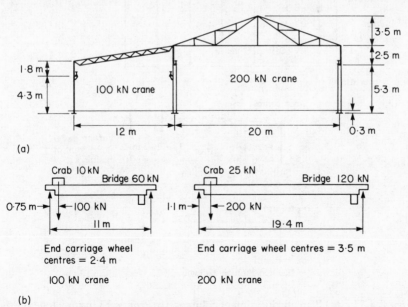

Figure 4.30   (a) Building cross-section. (b) Crane loads.

and sheeting rails. The building is located on the outskirts of a city. Carry out the following work.

(a) Draw the framing plans for the building.
(b) Estimate the loading on an internal frame.
(c) Analyse and design the roof trusses.
(d) Analyse and design the crane girders.
(e) Analyse and design the columns and bases.
(f) Make an arrangement drawing of the frame and show enlarged views of the main structural details.

3. The plan for an agricultural exhibition building is shown in Fig. 4.31(a). The front of the building is to be a clear opening free of columns. The clear height of the building is 5 m. Two proposals for framing the roof are shown in Fig. 4.31(b) and (c). Carry out the following work for each proposal.

(a) Draw complete framing plans showing all the bracing required.
(b) Make preliminary designs.
(c) Calculate the weight of steel required for each.

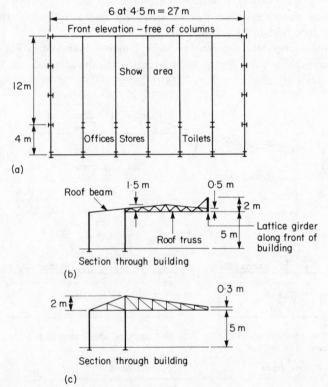

Figure 4.31   (a) Plan and location of columns. (b) Proposal 1, using a longitudinal girder. (c) Proposal 2, using cantilever trusses.

References and further reading

[1] BS449, Part 2 (1969). *The use of Structural Steel in Building*. British Standards Institution, London.

[2] CP3, Chapter V, Part 2 (1972). *Wind Loads*. British Standards Institution, London.

[3] *Steel Designers Manual* (1972). Constrado and Crosby Lockwood, London.

[4] *Structural Steelwork Handbook* (1978). B.C.S.A. and Constrado, London.

[5] MacGinley, T. J. (1973). *Structural Steelwork Calculations and Detailing*. Butterworths, London.

[6] CP3, Chapter V, Part 1 (1967). *Dead and imposed loads*. British Standards Institution, London.

5 Framed buildings—rigid elastic design

5.1 Types and general considerations

The rigid frame building is made with rigid joints capable of transmitting moment shear and axial force. There is no relative rotation between members meeting at joints. Rigid frame buildings may be classified into the following types:

single-storey portals   — single- or multi-bay;
multi-storey buildings — single- or multi-bay.

Examples of rigid plane frames are shown in Fig. 5.1(a) and (b). Rigid frames tend to be identified with plane frames because structures are idealized in this way for analysis and design. Buildings are three-dimensional and it is essential to take the whole structure into account when considering stability. Further considerations regarding this point and complete framing plans are given in the next section.

Rigid frames are made of rolled universal beams or structural hollow sections or built-up I or box sections. The structures are aesthetically attractive, clear of internal trusses and thus easy to protect against corrosion. It is difficult to say whether a rigid frame design will cost less than an equivalent simple design. Less

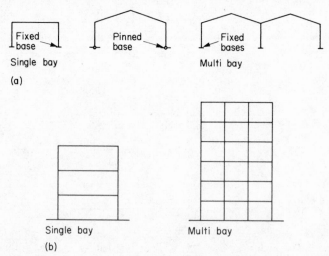

Figure 5.1   (a) Single-storey rigid frames. (b) Multi-storey frames.

steel may be used in the rigid design but the joints will be more expensive to make and the fabrication must be more accurate. A saving in the overall height of a building can often be made if rigid design is adopted and this saves material, maintenance cost and space to be heated.

5.2 Framing plans and typical joint details

The framing plans for a single-storey industrial building with crane are shown in Fig. 5.2(a). The building is rigid in the transverse direction but of simple construction longitudinally with bracing provided in the roof and walls in the end bays. Note that the longitudinal crane surge is picked by bracing.

Multi-storey buildings may be rigid in both directions but more often are rigid in one direction only, i.e. the direction in which the building loads are to be carried. This is because construction is greatly simplified and because I sections are commonly used. The sections are oriented so the much stronger major axis takes bending in the plane of the rigid frame. Stability in the other direction is provided by bracing, shear walls or service shafts. Rigid frames may be arranged and oriented to provide stability in both directions. See Section 6.2. A true space structure is possible using box members. The framing plans for a multi-storey building are shown in Fig. 5.2(b). The reader should also refer to Chapter 10 where the structural systems for high-rise buildings are discussed.

The location and classification of the commonly used joints in rigid frame construction are shown in Fig. 5.3(a) and typical details are shown in (a), (b) and (c). Welded fabrication and the use of site welding and high-strength friction bolts for the rigid joints are an essential feature of rigid frame construction. Haunches are often provided at joints to reduce stresses where moments are high. The effect of haunches on the analysis of the structure is considered in Section 5.3.3. Beam-to-column joints (see Fig. 5.3c) can be site welded but fabrication and erection have to be very accurate. In bolted joints, haunching is necessary to increase the lever arm of the bolt group to give a joint with high moment capacity.

5.3 Elastic analysis

5.3.1 Rigid frame analysis

The methods used for rigid frame analysis are discussed briefly. The reader is referred to text books on structural mechanics for a complete treatment [2, 3]. The main classification is into (a) force or flexibility method, and (b) displacement of stiffness method.

(a) *Force or flexibility method.* The structure is released to make it statically determinate and the displacements at the releases due to the frame loads are calculated. The redundants or the actions needed to restore continuity or cancel out displacements in the released structures are determined by setting

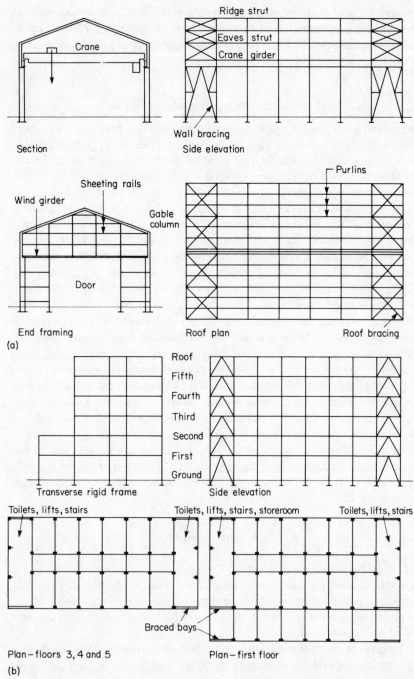

Section

Crane

Side elevation

Ridge strut

Eaves strut

Crane girder

Wall bracing

End framing

Wind girder

Sheeting rails

Door

Roof plan

Gable column

Purlins

Roof bracing

(a)

Transverse rigid frame

Side elevation

Roof
Fifth
Fourth
Third
Second
First
Ground

Toilets, lifts, stairs

Toilets, lifts, stairs, storeroom

Toilets, lifts, stairs

Braced bays

Plan – floors 3, 4 and 5

Plan – first floor

(b)

Figure 5.2 (a) Single-storey industrial building. (b) Multi-storey rigid frame building.

up and solving a set of simultaneous equations. The traditional methods are the strain energy, virtual work or area—moment methods. The method can be formulated using matrix algebra and programmed for specific applications.

(b) *Displacement of stiffness method.* Restraining forces are applied to prevent the displacement at the joints. These forces are determined as a sum of the fixed end forces for the members meeting at each joint. Each joint is now given a unit displacement in turn. The joints of the plane frame are subjected to three displacements while those of a space frame can have six displacements. The internal and external forces are then determined. The values of the displacements necessary to eliminate the restraining forces are determined by setting up and solving a set of simultaneous equations. The forces in the structure can then

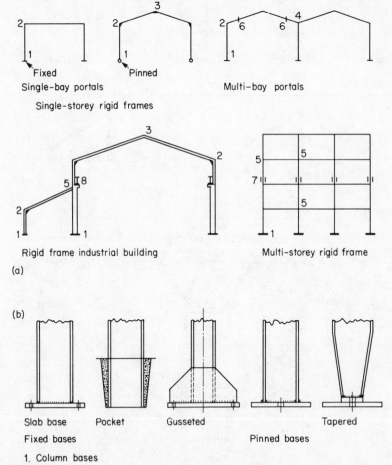

Figure 5.3 (a) Location of joints. 1 Base — pinned or fixed; 2 Knee joint; 3 Apex joint; 4 Valley joint; 5 Beam-to-column joint; 6 Rafter splice, beam splice; 7 Column splice; 8 Crane girder joint. (b) Types of joints.

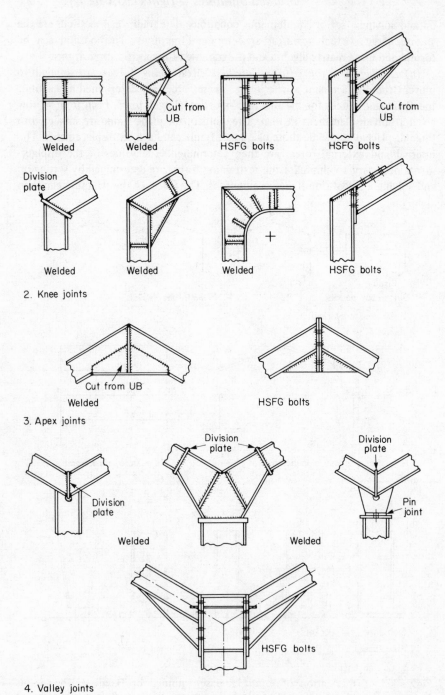

2. Knee joints

3. Apex joints

4. Valley joints

Figure 5.3(a)

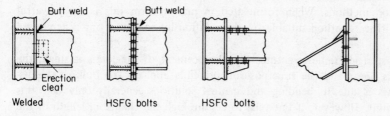

Welded   **HSFG bolts**   **HSFG bolts**

5. Beam–column connections

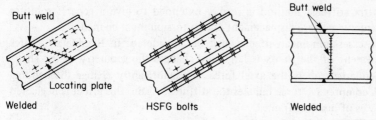

Welded   **HSFG bolts**   **Welded**

6. Rafter and beam splices

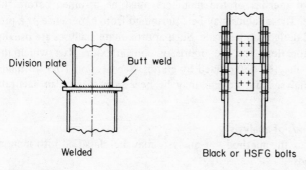

Welded   **Black or HSFG bolts**

7. Column splices

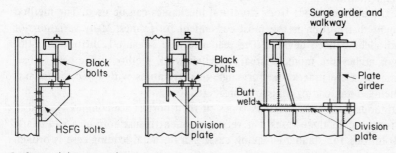

8. Crane girder connections

Figure 5.3(b)

be determined. The traditional methods are the slope deflection and moment distribution methods. When formulated in matrix form this is the powerful matrix stiffness method suitable for the solution by computer or a general rigid frame.

In most rigid frame structures the displacement of the joints is small, the analysis is linear and the superposition principle applies. The bulk of the displacements is due to bending and manual solutions generally only take this into account. However, if the axial forces are high the change in length of the members sets up secondary moments. The general matrix stiffness method takes this effect into account. Displacements due to shear force are not significant in building frames.

The matrix stiffness method is readily extended to give a general solution for space structures. Here displacement due to torsion must be included.

In some cases non-linear effects must be considered. In flexible structures the displacement of the joints is large and changes the geometry of the structure. In slender members the axial force can significantly change the bending stiffness. A compressive force reduces the stiffness. Again, this effect is neglected in the analysis of building frames.

For hyperstatic structures, the distribution of moments depends on the relative stiffness of the members. Thus the sizes giving the properties or the ratios of moments of inertias of the members must be assumed before the analysis can be made. These values may be determined from experience or a preliminary analysis and design can be made. Such approximate analyses are usually made by assuming locations for the points of contraflexure after which the moments and forces may be determined by statics. See Section 10.6. In many cases, several trial analyses and designs may be necessary to give an accurate result.

5.3.2 Practical methods of analysis
For practical purposes the methods of analysis may be classified into manual and computer methods.

(a) Manual methods
Any of the basic methods from structural mechanics can be used. The method of moment distribution is the most convenient for frames. Many variations of the method have been devised. The reader should consult Lightfoot [5]. For portals or arches the moment area or virtual work method is the best to use. These methods are practicable for single-cell structures with three redundants only when general loading cases are considered.

Continuous beams and various types of rigid portals are commonly used in steel structures. In these cases it is very convenient to use formulae and charts giving solutions for standard loading cases. Any general loading case is broken down into the separate cases for which solutions are given and then these results are combined by the principle of superposition. Solutions are given in the *Steel Designers Manual* [7].

(b) Computer programs

The computer analysis is based on the matrix stiffness method. Programs are readily available to analyse plane frames, grids and space frames. The programs are in two parts – input data and output. In detail these parts consist of:

Data   (i) joint co-ordinates referenced to the frame axes.

      (ii) member connection giving joints between which members lie and whether joints are pinned or fixed,

     (iii) assumed member properties – area, moments of inertia and torsion constant if required,

     (iv) restraints, i.e. support conditions pinned or fixed,

      (v) separate load cases. Point loads or moments may be applied at joints or any position along the member. Uniformly distributed or uniformly varying loads or moments may be applied over the whole or part of the member;

Output (i) deflections and joint rotations at all joints,

      (ii) axial forces, shears and moments at ends of each member,

     (iii) equilibrium checks.

The programs used in the book are the I.C.L. Plane Frame and Space Frame programs. The information for these are set out in the users manuals [4].

5.3.3 The effect of haunched joints

Most rigid frames, particularly single-storey portals, are constructed with haunches at the joints. If the haunches are small in comparison with the uniform part of the member, it is customary to analyse the frame for uniform members and design for the maximum moment at the joints. See Fig. 5.4(a). The haunches are provided to reduce the stresses at the points of maximum moment and with bolted joints provide the extra lever arm required to make the rigid joint.

A more economic design is achieved if the prismatic members are designed for the moment at the ends of the haunch. The haunch must then be designed by the methods set out in Section 5.4.2.

For an accurate solution, the haunches are taken into account in the analyses. This must be done where the haunches are large in comparison with the straight parts of members or where the frame is constructed with tapered members as shown in Fig. 5.4(b). The haunch attracts moment to itself so there is a higher moment at a haunched joint than at the same joint if the frame is analysed with uniform members.

The haunch is taken into account by dividing it into a number of sections. The properties over each section are assumed to be uniform. See Fig. 5.4(c). The average properties at the mid section are used in the computer analysis. The standard analysis programs are based on prismatic members. Computer programs can be written to include uniformly tapered members.

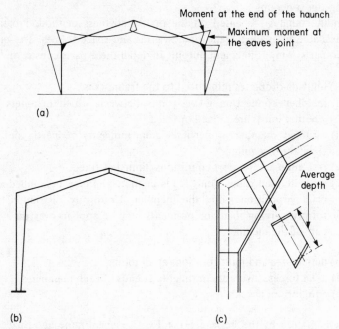

Figure 5.4 (a) Rigid portal — bending moment diagram (b) Frame with tapered members. (c) Division of the haunched joint.

5.4 Elastic design

In general, elastic design of members in rigid frames is the same as for those to simple design. However, the design of columns, members of portal frames and haunched joints requires further consideration.

5.4.1 Design of columns

One particular area of uncertainty in the design of columns in rigid frame buildings lies in determining the effective length to use to obtain the permissible compressive and bending stresses. Otherwise the design follows the principles outlined in Chapter 3.

For buildings where sway is prevented by braced bays or shear walls, no undue problems occur in estimating the effective length of columns. The recommendation set out in BS449, Clause 31 and Appendix D [11], can be used. The effective length is less than the actual length.

A more accurate determination can be made by using the Chart from Wood [9]. This is reproduced here in Fig. 5.5(a) by kind permission of the Building Research Establishment. For a complete treatment of the subject the reader should consult the reference.

Referring to Fig. 5.5(a), the ratios of column stiffness to total stiffness are

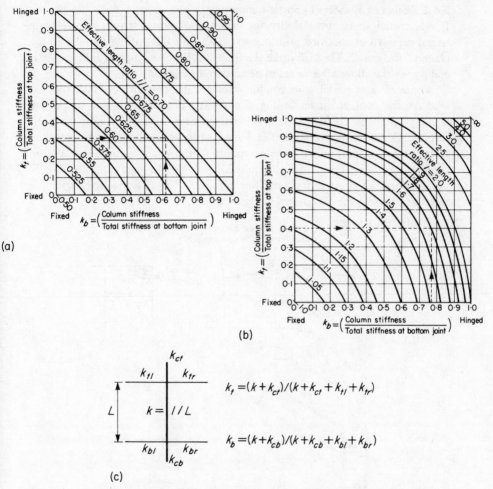

Figure 5.5   (a) Effective length ratios – no sway. (b) Effective length ratios – no restraint against sway. The charts are reproduced by kind permission from: Wood, R. W., *Effective length of columns in multistorey buildings*. Building Research Establishment.

determined at the top and bottom of the column. The effective length ratio can be read from the chart.

The stiffness K = Moment of intertia I/length centre to centre of beams L.

In rigid jointed buildings where side sway is not prevented, the effective length is greater than the actual length and can reach large values if the beams are flexible. The appropriate chart from the paper quoted above is very useful in these cases to give an accurate value. This chart is reproduced here in Fig. 5.5(b) with kind permission of the Building Research Establishment.

134   *Steel structures*

5.4.2 Design of members in portal frames

It is essential to ensure stability for members of portal frames by providing lateral supports at specified positions to prevent buckling at right angles to the plane of the portal. This will make the frame act in accordance with the analysis and permit the allowable stresses to be accurately assessed.

A pinned base portal with bending moment diagram for dead and imposed load on the roof is shown in Fig. 5.6(a). The members require supporting laterally on the inside flange which is in compression at the eaves, near the point of contraflexure in the rafter and in some cases near mid height of the column. Near the ridge, the purlins provide support to the compression flange and no support for the bottom flange is required. Uplift due to wind does not usually cause reversal of stress in this area. See Fig. 5.15. The effective lengths for the rafter and column are also shown in Fig. 5.6(a).

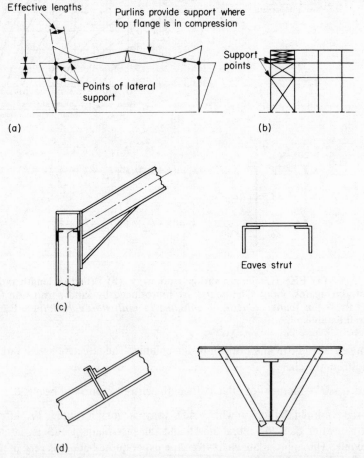

Figure 5.6 (a) Portal – bending moment diagram and support points. (b) Side bracing. (c) Eaves support member. (d) Bracing off purlins and sheeting rails.

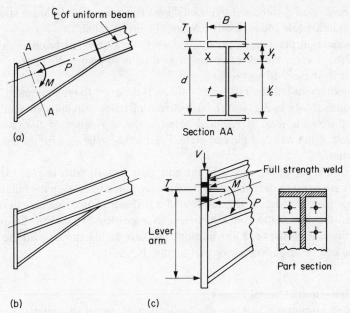

Figure 5.7   (a) Tapered beam. (b) Haunch cut from a universal beam. (c) Joint.

Special support members may be provided as shown in Fig. 5.6(c). Here the member consists of two angles battened together. Alternatively, the inside flange may be supported from purlins or sheeting rails as shown in Fig. 5.6(d). These support points should be picked up by longitudinal bracing in the end bays as shown in Fig. 5.6(b). However, it is usually assumed that the diaphragm action of the sheeting alone will provide sufficient restraint.

For buckling of members in the plane of the truss, the effective lengths can be found approximately by using Fig. 5.5(b). Alternatively, the effective lengths can be estimated using the guidance given in BS449 [11].

5.4.3 Design of haunched joints
Various methods are available for the determination of stresses in haunched joints (see [1, 7]). Vierendeel's tapered beam method is as follows: see Fig. 5.7(a).

y_t, y_c = distances from the centroids of the top and bottom flanges respectively to the XX axis

$A = BT + BT \cos \phi + dt$

$I = BTy_t^2 + BT \cos \phi \, y_c^2 + td^3/12$

f_c = maximum stress in the bottom flange

  $= P/A + My_c/I,$

where the section is subjected to an axial load P and moment M. Shear stresses in light portals are low and normally need not be investigated.

It is common practice to cut the haunch from a universal beam as shown in Fig. 5.7(b). Here a conservative design method is to ignore the interior flanges and treat the haunched section as in Fig. 5.7.

The joint between the rafter and column is designed to resist moment, thrust and shear as shown in Fig. 5.7(c). High-strength friction grip bolts are used.

A full-strength weld is required between the top flange of the rafter, the upper part of the web and the end plate. Fillet weld can be used for the remainder of the joint.

The design for the thickness of the end plate is dealt with in [12, 15]. Here methods are given to determine the thickness required to develop the full strength of the member connected. The addition of a stiffener reduces bending in the end plate and distributes the bolt load over a wide portion of the web. Simplifying assumptions as to the load distribution and plate action can be made to determine the end plate thickness. See part section Fig. 5.7(c).

5.4.4 Deflection of portal frames

It is often important to know the deflection at points on portals. Excessive deflection at the eaves may damage brick side walls and cause problems in buildings with cranes. The actual deflection is always less than that calculated for the steel frame only. This is due to the stressed skin action in the cladding. BS449, Clause 31b [11] limits the deflection at the stanchion cap due to lateral loads to 1/325 of the stanchion height, except in cases where greater deflection will not impair strength or efficiency or lead to damage to finishes. The allowable deflection in portals is a matter of engineering judgement.

If a computer analysis is used, the deflections are printed out as part of the output and are available for all joints for all load cases. For manual calculations the formulae for the horizontal deflection at the eaves due to a uniform roof load for pinned-base and fixed-base portals are given below. These have been derived using the moment area theorems.

(a) *Fixed-base portal*   See Fig. 5.8(a) which shows the dimensions, loading, moments and deflected shape for the structure. The deflection of the column top B away from the tangent at the base A is

$$\delta_B = h^2 \, (M_A - M_B/2)/EI_{AB}$$

where I_{AB} is the moment of intertia of the column AB, M_A the moment at the column base, and M_B the moment at the column top.

(b) *Pinned-base portal*   See Fig. 5.8(b). Because of symmetry the slope at the ridge does not change. The slope at the base is equal to the area of the M_{EI} diagram between the ridge and base. This is

$$\phi = [(wl^2 s/24) + M_B s + (Hbs/2) - (Vls/4)]/EI_{BC} + (M_B h/2EI_{AB}),$$

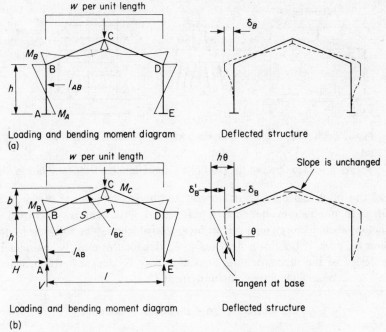

Figure 5.8   (a) Fixed-based portal. (b) Pinned-base portal.

where I_{BC} is the moment of inertia of the rafter BC, H the horizontal reaction at the base, V the vertical reaction at the base, and w the roof load.

Then the deflection at the eaves is

$$\delta_B = h\theta - M_B h^2 / 6EI_{AB}.$$

5.5 Foundations for rigid frame buildings

Building to simple design can accommodate some movement without over-stressing the members of the structure. Great care is required with the design of foundations for rigid frame buildings. The rotation of a base taken as fixed or differential settlement or horizontal displacement can cause a radical redistribution of moments in the frame resulting in overstressing. The structure may be analysed for a given displacement of a foundation and designed to take account of this. However, it is usual to design the foundations so that movement does not occur. Some basic cases are discussed.

Portals generate large horizontal thrusts at the bases due to loads on the rafters. This can cause isolated bases to spread. In many cases a tie can be put through the slab or in a ground beam to take the thrust. The foundations are then designed for axial load. Wind load must be taken by passive resistance and adhesion between the base and the soil. See Fig. 5.9(a). If isolated bases are necessary and ground conditions are poor, pinned bases should be used. Piled

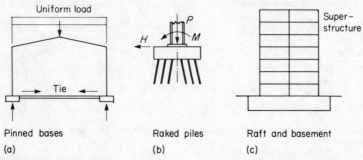

Figure 5.9   (a) Ground tie. (b) Piled foundation. (c) Raft foundation.

bases can be designed to resist moment and horizontal loads as shown in Fig. 5.9(b). In multi-storey buildings if differential settlement is anticipated the structure should be set on a raft or basement which may be supported on the soil or on piles. If the raft is designed to be flexible this must be considered in the design of the superstructure. Structures may be designed with articulated joints to accommodate foundation movement.

5.6 Elastic design — single-storey building

5.6.1 Specification

The section through an industrial building where the rigid frame is a uniform two-pinned portal is shown in Fig. 5.10. The frames are at 5 m centres and the length of the building is 40 m. The building loads are:

Roof — dead load measured on slope	(kN/m^2)
sheeting 20 gauge, purlins at 1.75 centres	0.1
insulation board	0.15
purlins 120 x 120 x 8L design to BS449, Clause 45	0.1
portal rafter, assume 457 x 191 x 67 kg/m UB	0.13
Total	0.48

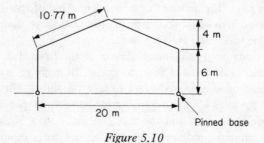

Figure 5.10

— imposed load measured on plan CP3, Chapter V, Part 1   0.75 kN/m^2

Walls — sheeting, insulation board, sheeting rails, stanchion   0.48 kN/m^2

Wind loading — CP3, Chapter V, Part 2. The location of the building is in the North East of England on the outskirts of a city.
Carry out the following work:

(a) Estimate the loading on an internal portal due to dead and imposed load and wind internal pressure and internal suction cases with the wind transverse to the building and the maximum uplift case when the wind is blowing longitudinally.
(b) Analyse the portal for the five separate load cases listed in (b) and draw the bending moment diagrams for each case.
(c) Design the portal using Grade 43 steel.

5.6.2 Loading

(a) Dead and imposed loads
Roof — the loading is considered to be uniformly distributed on plan

dead load $= 5 \times 0.48 \times 10.77/10 =$ 2.48 kN/m

imposed load$= 5 \times 0.75$ $=$ 3.75 kN/m

Walls — dead load $= 0.48 \times 6 \times 5$ $= 14.4$ kN

The loading is shown in Fig. 5.12(a) and (b).

(b) Wind loads

Basic wind speed V =   45 m/s

Topography factor $S_1 = 1.0$ and statistical factor $S_3 = 1.0$

Ground roughness is category 3 and building size is Class B.

S_2 — factors from Table 3, CP3, Chapter V, Part 2 [14], the design wind speeds and dynamic pressures for the roof and walls are given in Table 5.1.

The external pressure coefficients C_{pe} from Table 8, CP3, Chapter V, Part 2, are shown in Fig. 5.11.

Table 5.1   Wind loads

	Height, H (m)	Factor S_2	Design wind speed $V_s = S_2 V$ (m/s)	Dynamic pressure $q = 0.613 V_s^2$ (N/m^2)
Roof	10	0.74	33.3	682
Wall	6	0.67	30.2	558

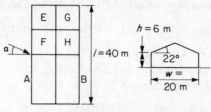

Figure 5.11

The internal pressure coefficients C_{pi} from Appendix E, CP3, Chapter V, Part 2, for the case where there is a negligible probability of a dominant opening occurring during a severe storm are +0.2 or –0.3. The three wind cases examined are:

(i) Wind angle 0° – internal pressure case
(ii) Wind angle 0° – internal suction case
(iii) Wind angle 90° – internal pressure case – maximum uplift.

The wind load on any particular surface is given by:

$$w = 5q(C_{pe} - C_{pi})/10^3 \text{ kN/m.}$$

The pressure coefficients and wind loads are shown in Fig. 5.12 (c), (d) and (e) for the three cases listed above.

5.6.3 Analyses

(a) Unit load cases
The loading on the frame can be broken down into three separate load cases such that each actual load case can be built up from these three basic cases. The unit load cases are shown in Fig. 5.13(a).

The analysis is made for a frame of uniform section ignoring the effect of haunches. The solutions for the three load cases to be analysed are taken from the *Steel Designers Manual* [7]. The frame data for the pinned-based portal are shown in Fig. 5.13(b). The loading and formulae, values for the moments and reactions and the bending moment diagrams for the three load cases are shown in Fig. 5.13(c), (d) and (e), respectively.

The bending moment diagrams for the separate load cases can now be formed from the unit load cases. The working for the wind load internal pressure case is shown in Fig. 5.14 to demonstrate the method. The normal wind loads on the roof slopes are resolved vertically and horizontally in order that the solutions for the unit cases can be used. The bending moment diagrams for the separate load cases are shown in Fig. 5.15.

For the dead load, the moment at any point P in the rafter is given by (see Fig. 5.15a) $M_p = 20.89x - 58.52 - 1.24x^2$.

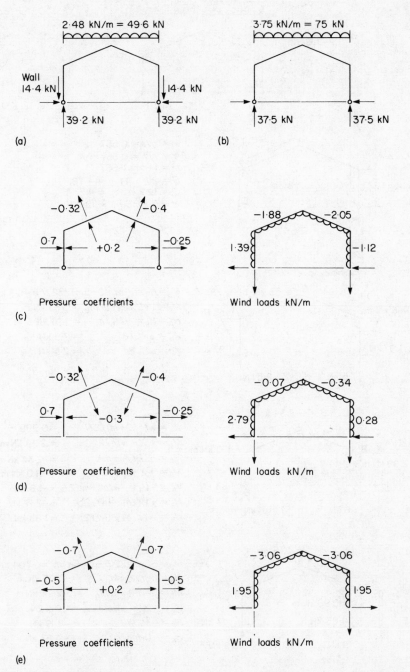

Figure 5.12 (a) Dead load. (b) Imposed load. (c) Wind angle 0° – internal pressure case. (d) Wind angle 0° – internal suction case. (e) Wind angle 90° – internal pressure case.

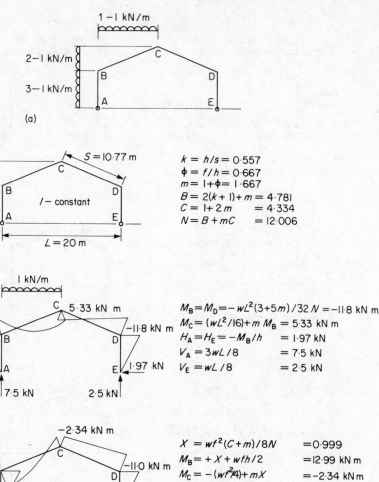

(a)

(b)

$S = 10.77$ m
$f = 4$ m
$h = 6$ m
l — constant
$L = 20$ m

$k = h/s = 0.557$
$\phi = f/h = 0.667$
$m = 1+\phi = 1.667$
$B = 2(k+1)+m = 4.781$
$C = 1+2m \quad = 4.334$
$N = B+mC \quad = 12.006$

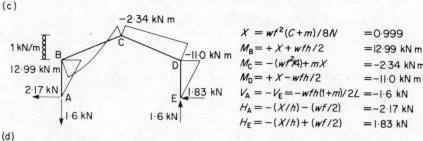

(c)

1 kN/m

C 5.33 kN m
−11.8 kN m
−11.8 kN m
1.97 kN
1.97 kN
7.5 kN
2.5 kN

$M_B = M_D = -wL^2(3+5m)/32N = -11.8$ kN m
$M_C = (wL^2/16)+m\,M_B = 5.33$ kN m
$H_A = H_E = -M_B/h \quad = 1.97$ kN
$V_A = 3wL/8 \quad = 7.5$ kN
$V_E = wL/8 \quad = 2.5$ kN

(d)

−2.34 kN m
1 kN/m
12.99 kN m
2.17 kN
−11.0 kN m
1.83 kN
1.6 kN
1.6 kN

$X = wf^2(C+m)/8N \quad =0.999$
$M_B = +X+wfh/2 \quad =12.99$ kN m
$M_C = -(wf^2/4)+mX \quad =-2.34$ kN m
$M_D = +X-wfh/2 \quad =-11.0$ kN m
$V_A = -V_E = -wfh(1+m)/2L = -1.6$ kN
$H_A = -(X/h)-(wf/2) \quad =-2.17$ kN
$H_E = -(X/h)+(wf/2) \quad =1.83$ kN

(e)

−2.74 kN m
1 kN/m
10.96 kN m
4.83 kN
−7.04 kN m
1.17 kN
0.9 kN
0.9 kN

$M_D = -(wh^2/8)(2(B+C)+K)/N = -7.04$ kN m
$M_B = (wh^2/2)+M_D \quad = 10.96$ kN m
$M_C = (wh^2/4)+m\,M_D \quad = -2.74$ kN m
$V_A = -V_E = -wh^2/2L = -0.9$ kN
$H_E = -M_D/h \quad = -1.17$ kN
$H_A = -(wh-H_E) \quad = -4.83$ kN

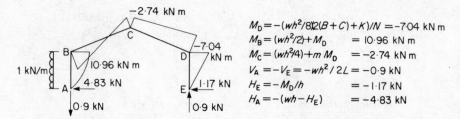

Figure 5.13 (a) Unit load cases. (b) Frame data. (c) Case 1, 1 kN/m vertical load on BC. (d) Case 2, 1 kN/m horizontal load on BC. (e) Case 3, 1 kN/m horizontal load on AB.

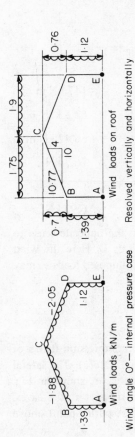

Wind loads kN/m

Wind angle 0° — internal pressure case

Wind loads on roof

Resolved vertically and horizontally

Action	Horizontal load on AB →		Horizontal load on BC ←		Vertical load on BC ↑		Vertical load on CD ↑		Horizontal load on CD →		Horizontal load on DE →		Totals Moments Reactions
	1.0 kN/m	1.39 kN/m	1.0 kN/m	0.7 kN/m	1.0 kN/m	1.75 kN/m	1.0 kN/m	1.9 kN/m	1.0 kN/m	0.76 kN/m	1.0 kN/m	1.12 kN/m	
M_B	10.96	15.23	−12.99	−9.09	11.8	20.65	11.88	22.42	11.0	8.36	7.04	7.88	66.45
M_C	− 2.74	− 3.81	2.34	1.64	− 5.33	− 9.33	− 5.33	−10.13	2.34	1.78	2.74	3.07	−16.78
M_D	− 7.04	− 9.78	11.00	7.7	11.8	20.65	11.8	22.42	−12.99	−9.87	−10.96	−12.28	18.84
H_A	− 4.83	− 6.71	2.17	1.52	− 1.97	− 3.45	− 1.97	− 3.74	− 1.83	−1.39	− 1.17	− 1.31	−15.08
H_E	− 1.17	− 1.63	1.83	1.28	1.97	3.45	1.97	3.74	− 2.17	−1.65	− 4.83	− 5.41	− 0.22
V_A	− 0.9	− 1.25	1.6	1.12	− 7.5	−13.13	− 2.5	− 4.75	− 1.6	−1.22	− 0.9	− 1.01	−20.24
V_E	0.9	1.25	− 1.6	− 1.12	− 2.5	− 4.39	− 7.5	−14.25	1.6	1.22	0.9	1.01	−16.27

Sign covention: Moments causing tensions on inside of frame +ve

Horizontal reactions → +ve

Vertical reactions ↑ +ve

Units Moments kN m

Reactions kN

Figure 5.14

Putting $M_P = 0$ and solving gives $x = 3.55$ m, the location of the point of contraflexure in the rafter BC.

Putting $dM_P/dx = 0$ and solving gives $x = 8.26$ m, the location of the maximum sagging moment in the rafter. The value of this moment is 29.47 kN/m.

For the imposed load case the value of this moment is 44.56 kN m.

5.6.4 Frame design

(a) Design data

The design data from the bending moment diagrams on Fig. 5.15 are:

Dead + imposed loads

$$\text{Eaves} \quad M_D = -58.52 - 88.48 \qquad\qquad = -147 \text{ kN m}$$

$$P_D = \quad 24.8 + 37.5 \qquad\qquad = \quad 62.3 \text{ kN}$$

$$\text{Ridge} \quad M_C = \quad 26.44 + 39.98 \qquad\qquad = \quad 66.42 \text{ kN m}$$

$$\text{Base} \quad P_E = \quad 39.2 + 27.5 \qquad\qquad = \quad 76.7 \text{ kN}$$

Dead + wind

$$M_D = -58.52 + 65.45 \qquad\qquad = \quad 6.93 \text{ kN m}$$

Dead + imposed + wind – the actions due to wind are in the opposite sense to those due to dead and imposed loads except at the eaves D. Here the wind moment does not exceed 25% of moment due to dead and imposed loads so may be neglected.

Other actions required will be calculated in the design.

(b) Lateral support for the frame

To reduce the effective length of the stanchion and the compression flange of the rafter, lateral supports are provided as shown in Fig. 5.16(a). The lateral supports at mid height of the column and in the rafter area are shown in Fig. 5.16(b). A battened member is provided to give lateral support at the eaves. The bottom flange of the rafter is supported at the first purlin point above the eaves. This is about 1.9 m from the eaves.

(c) Design of the side column

Assume 457 x 152 UB 60. The properties are:

$$A = 75.9 \text{ cm}^2 \qquad\qquad r_{YY} = 3.23 \text{ cm}$$

$$Z_{XX} = 1120 \quad \text{cm}^3 \qquad\qquad D/T = 34.2$$

$$r_{XX} = 18.3 \text{ cm} \qquad\qquad I_{XX} = 25\ 464 \text{ cm}^4$$

The effective length for the XX-axis l_{XX} is estimated using Fig. 5.5(b)

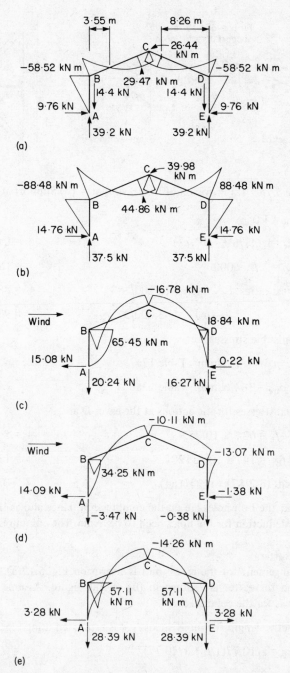

(a)

(b)

(c)

(d)

(e)

Figure 5.15 Bending moment diagrams and reactions. (a) Dead loads. (b) Imposed loads. (c) Wind angle $0°$ – internal pressure case. (d) Wind angle $0°$ – internal suction case. (e) Wind angle $90°$ – internal pressure case.

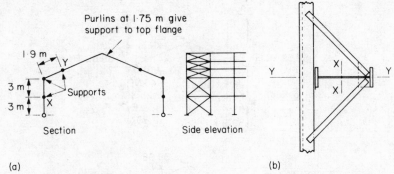

Figure 5.16   (a) Provision of lateral supports. (b) Support at X and Y.

Here $k_b = 1.0$

$\qquad k_t = 1/6(1/6 + 1/10.77)^{-1}$ $\hfill = 0.64$

$\qquad l_{XX} = 2.7 \times 6000$ $\hfill = 16\,200 \quad$ mm

Slenderness ratios: $l_{XX}/r_{XX} = 16\,200/183$ $\hfill = 88.5$

$\qquad\qquad\qquad l_{YY}/r_{YY} = 3\,000/32.3$ $\hfill = 93$

The permissible stresses are:

Axial $\qquad p_c = 87$ N/mm^2 Table 17a

Bending $p_{bc} = 164$ N/mm^2 Table 3a

The actual stresses for the actions at the eaves D are:

Axial $\qquad f_c = 62.3 \times 10/75.9$ $\hfill = 8.2$ N/mm^2

Bending $f_{bc} = 147 \times 10^3/1120$ $\hfill = 131$ N/mm^2

Combined: $(8.2/87) + (131/164)$ $\hfill = 0.895$ Safe.

Note that the bolt holes are on the compression flange and as HSFG bolts are used no deduction for the holes need to be made. The column is satisfactory.

(d) Portal Rafter

The arrangement for the eaves joint is shown on Fig. 5.17(a). The shear and thrust on the rafter are shown in (b) on the figure. Assume the same size member as for the column.

The effective length for the XX-axis l_{XX} is estimated using Fig. 5.5(b)

Here $\quad k_b = 1/10.77(1/6 + 1/10.77)^{-1}$ $\hfill = 0.36$

$\qquad k_t = 0.5$

$\qquad l_{XX} = 1.36 \times 10\,770$ $\hfill = 14\,647$ mm

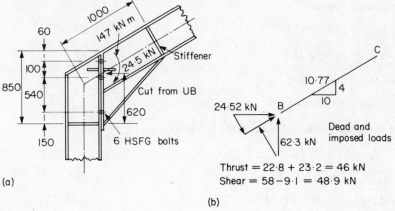

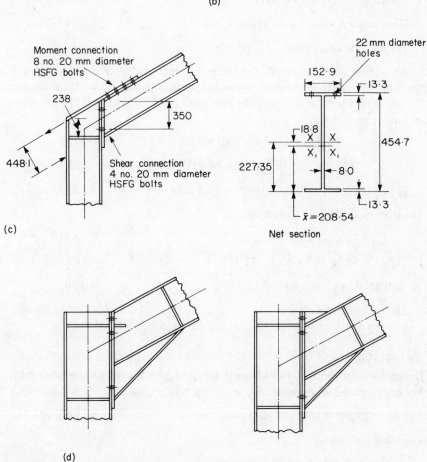

Figure 5.17 (a) Joint. (b) Thrust and shear in rafter BC. (c) Alternative arrangement for joint. (d) Other designs for the joints.

Slenderness ratios

$l_{XX}/r_{XX} = 14\ 647/183$ $= 80$

$l_{YY}/r_{YY} =\ \ 1\ 900/32.3$ $= 59$

$\qquad p_{bc} = 165\ \text{N/mm}^2$

$\qquad\quad p_c = 104\ \text{N/mm}^2$

The stresses are checked neglecting the haunch

$\qquad f_c\ = 46 \times 10/75.9$ $= 6.1\ \text{N/mm}^2$

$\qquad f_{bc} = 147 \times 10^3/1120$ $= 131\ \text{N/mm}^2.$

Combined $(6.1/104) + (131/165)$ $= 0.853.$ Safe.

Shear stress $f_s = 48.9 \times 10^3/(454.7 \times 8)$ $= 13.4\ \text{N/mm}^2.$

The rafter is satisfactory.

An alternative arrangement for the eaves joint is shown in Fig. 5.17(c). In this case the rafter is checked on the net section shown on the figure because the top flange is in tension and the bolt holes must be deducted. The properties of the net section are:

$\qquad A\ = 75.9 - (2 \times 22 \times 13.3/100) = 75.9 - 5.85$ $= 70.05\ \text{cm}^2$

$\qquad x\ = [(75.9 \times 22.74) - (5.85 \times 44.81)]/70.05$ $= 20.85\ \text{cm}$

$\qquad I_{X_1 X_1} = 25\ 464 + 70.05 \times 1.88^2 - 5.85 \times 23.96^2$ $= 22\ 353\ \text{cm}^4.$

The stresses at the next section are:

$\qquad f_c\ = 46 \times 10/70.05$ $= 6.6\ \text{N/mm}^2$

$\qquad f_{bc}\ = 147 \times 10^3 \times 20.85/22\ 353$ $= 137.1\ \text{N/mm}^2$

Combined $(6.6/104) + (137.1/165)$ $= 0.89.$

The rafter is satisfactory.
Further arrangements for the joint are shown in Fig. 5.17(d).

(e) Eaves joint
Design for a moment of 147 kN m, a thrust of 24.5 kN and a shear of 62.3 kN. For the proposed arrangement shown in Fig. 5.17(a), the design is as follows:

$\Sigma y^2 = 2(150^2 + 690^2 + 790^2)$ $= 22.4 \times 10^5.$

Tension in top bolts:

$$T = \frac{[147 - 24.5 \times 0.62]10^3 \times 790}{22.4 \times 10^5}$$ $= 46.4\ \text{kN}.$

Shear assuming uniform distribution:

$S = 62.3/6$ $= 10.4$ kN.

Assume 20 mm HSFG bolts;   proof load $= 144$ kN.

Reduced proof load $= 144 - (1.7 \times 46.4)$ $= 65.1$ kN.

Shear value $= 65.1 \times 0.45/1.4$ $= 20.9$ kN/bolt.

The joint is satisfactory using 20 mm HSFG bolts. The joint is also satisfactory when checked for the reversed moment of 6.93 kN m due to dead and wind loads.

The design of the joint for the alternative arrangement shown in Fig. 5.17(c) is as follows:

Top bolts – Shear due to moment $= [147 - 24.5 \times 0.238]/0.448 = 315.1$ kN. Provide 8 no. 20 mm diameter HSFG bolts where the single shear value per bolt is 46.3 kN. The vertical shear of 48.9 kN will be resisted by 4 bolts, in the web joint.

Check the bottom bolts for the reversed moment due to wind of 6.93 kN m. Neglect the thrust.

Bolt tension $= 6.93/(2 \times 0.35)$ $= 9.9$ kN.

The bolts are satisfactory.

(f) Ridge joint

The proposed arrangement for the ridge joint is shown in Fig. 5.18. The joint is designed for a moment of 66.4 kN m. The thrust is neglected.

$\Sigma y^2 = 2[100^2 + 390^2 + 570^2]$ $= 9.74 \times 10^5$.

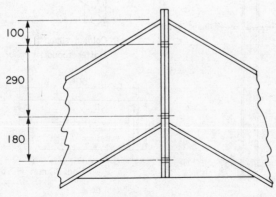

Figure 5.18

Tension in bottom bolts:

$$T = \frac{66.4 \times 10^3 \times 570}{9.74 \times 10^5} \qquad\qquad = 39 \text{ kN.}$$

Use six no. 20 mm HSFG bolts.

(g) Drawing
A sketch of the portal is shown in Fig. 5.19.

(h) Deflection of the rafter at the eaves
The horizontal deflection of portal at the eaves due to dead and imposed load is
calculated. See Section 5.4.4.

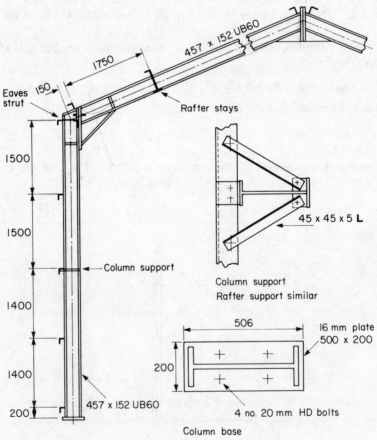

Figure 5.19

The slope at the base of the portal is given by:

$$\phi = [(6.23 \times 10\,000^2 \times 10\,770)/24 + (147 \times 10^6 \times 10\,770) +$$

$$+ (24\,520 \times 10\,770/2) - (62\,300 \times 10\,000 \times 10\,770/4)$$

$$+ (147 \times 10^6 \times 6\,000/2)]/(2.1 \times 10^5 \times 25\,464 \times 10^4)$$

$$= 1.71 \times 10^{-2} \text{ radians.}$$

The deflection at the eaves:

$$\delta_B = 6000 \times 1.71 \times 10^{-2} - \frac{147 \times 10^6 \times 6\,000^2}{6 \times 2.1 \times 10^5 \times 25\,464 \times 10^4}$$

$$= 70.29 - 16.49$$

$$= 53.8 \text{ mm.}$$

The part due to dead load 21.4 mm can be offset in the fabrication. The deflection due to imposed load is 32.4 mm equal to 1/185 of the column height.

5.6.5 Redesign of the portal rafter with a haunch

Assume the rafter section is 406 x 140 UB 39. The column section 457 x 152 UB 60 is unchanged. The frame is re-analysed with these sections. This gives:

$M_B = M_D$	$= -154.36$ kN m
M_C	$= 54.18$ kN m
H	$= 25.73$ kN.

The moment at any point P at x from B is:

$$M_P = 3.115x^2 - 62.3x + 25.73 \times 0.4x + 154.36$$

$$= 3.11x^2 - 52.01x + 154.36.$$

Put $dM_P/dx = 0$ and solve to give $x = 8.35$ m.

Maximum sagging moment M_F	$= 62.82$ kN m.
Thrust at F: P_F	$= 27.7$ kN.

The bending moment diagram for the rafter is shown in Fig. 5.20(a).

The column is checked for the new design actions. See Section 5.6.4(c).

$f_c = 8.2$ N/mm^2	
$f_{bc} = 154.36 \times 10^3/1120$	$= 137.8$ N/mm^2
Combined $(8.2/87 + (137.8/164)$	$= 0.94$

The section is satisfactory.

Ideally, the rafter is designed for the maximum sagging moment due to dead

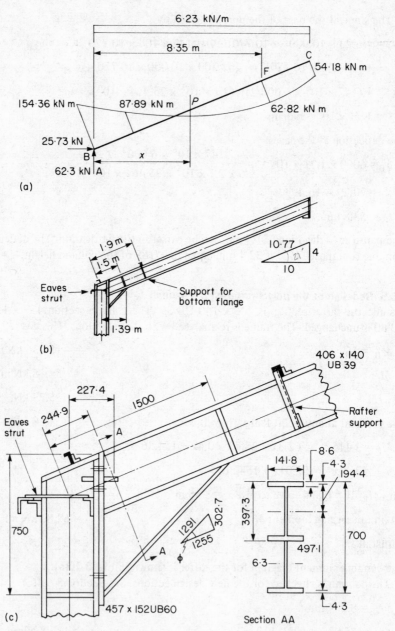

Figure 5.20 (a) Bending moment diagram for the rafter. (b) Portal rafter. (c) Haunched joint.

and imposed load. The haunch is made long enough so that at its end the hogging moment is reduced sufficiently to make the section safe. The assumed section for the rafter is 406 x 140 UB 39 where:

$$A = 49.4 \text{ cm}^2, \qquad r_{XX} = 15.88 \text{ cm}, \qquad I_{XX} = 12\ 452 \text{ cm}^4$$

$$Z_{XX} = 626.9 \text{ cm}^3, \qquad r_{YY} = 2.89 \text{ cm}.$$

At F the position of the maximum sagging moment, the top flange is in compression and is restrained by the purlins at 1.75 m centres. The effective length ratio for the rafter for buckling about the XX axis is estimated using Fig. 5.5(b).

$l/r_{XX} = 1.3 \times 10\ 770/158.8$ $\qquad = 88.1$

$l/r_{YY} = 1\ 750/28.9$ $\qquad = 61$

$p_c = 94 \text{ N/mm}^2$

$p_{bc} = 165 \text{ N/mm}^2$

$f_c = 27.7 \times 10/49.4$ $\qquad = 5.61 \text{ N/mm}^2$

$f_{bc} = 62.82 \times 10^3/626.4$ $\qquad = 100.3 \text{ N/mm}^2$

Combined $(5.61/94) + (100.3/165)$ $\qquad = 0.67$

The section is satisfactory. A lighter section could be used. Try a haunch 1.5 m long as shown in Fig. 5.20(b). Here:

Moment $M_H = 87.89 \text{ kN m}$

Thrust $P_H = 43.8 \text{ kN}$.

The bottom flange is in compression and supports to the bottom flange are provided at the second purlin above the eaves.

$l/r_{YY} = 1900/28.9$ $\qquad = 65.7$

$p_c = 94 \text{ N/mm}^2$ (from l/r_{XX} value above)

$p_{bc} = 165 \text{ N/mm}^2$

$f_c = 43.8 \times 10/49.4$ $\qquad = 8.87 \text{ N/mm}^2$

$f_{bc} = 87.89 \times 10^3/626.4$ $\qquad = 140.3 \text{ N/mm}^2$

Combined $(8.87/94) + (140.3/165)$ $\quad = 0.94$
The section is satisfactory.

The haunch is checked at Section AA shown in Fig. 5.20(c), using Vierendeel's tapered beam theory given in Section 5.4.3. The flanges on the original beam are ignored. These are taken as stiffening the web.

Here the design actions are:

Moment M = 142.7 kN m

Thrust P = 46.33 kN

$A = (141.8 \times 8.6) + (141.8 \times 8.6 \times 1255/1291) + (691.4 \times 6.3)$

$\quad = 1219 + 1185 + 4356 \qquad\qquad\qquad\qquad = 6760 \text{ mm}^2$

$I = (1219 \times 194.4^2) + (1185 \times 497.1^2) + 4356 \times 691.4^2/12$

$$= 5.12 \times 10^8 \text{ mm}^4$$

Maximum stress $f_c = \dfrac{46.33 \times 10^3}{6760} + \dfrac{142.7 \times 10^6 \times 497.1}{5.12 \times 10^8}$

$\qquad\qquad\qquad = 6.85 + 138.5 \qquad\qquad\qquad = 145.35 \text{ N/mm}^2$

The haunch is satisfactory.

A lighter section for the rafter coupled with a longer haunch could not be used. This is because the haunch would be overstressed when the next lighter section is used.

5.7 Coursework exercises

1. The steel frame for a warehouse is shown in Fig. 5.21. The building is 42 m long and the frames are at 6 m centres. The building is clad with plastic-coated corrugated steel sheet on insulation board. The loading is as follows:

Roof dead load – sheeting, insulation board,
purlins and rafters $\qquad\qquad\qquad\qquad\qquad$ = 0.5 kN/m^2

Roof imposed load – on plan $\qquad\qquad\qquad\qquad$ = 0.75 kN/m

Walls, dead load – sheeting, insulation board,
sheeting rails column $\qquad\qquad\qquad\qquad\qquad$ = 0.5 kN/m^2

Wind load CP3, Chapter V, Part 2 [14].

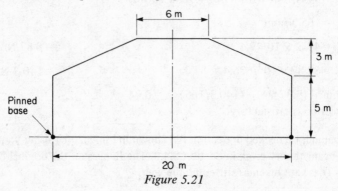

Figure 5.21

Design work required:

(a) Estimate the loading on an internal frame for the separate load cases of dead, imposed, wind load internal pressure and wind load internal suction. Show the loads on sketches.
(b) Analyse the portal for the separate load cases and draw the bending moment diagrams for each load case.
(c) Design the frame and show the main details on a drawing.
(d) Draw the framing plans for the building.

2. The framing plans for an office building and ground-floor showroom are shown in Fig. 5.22. The upper storeys are supported on rectangular pinned-base portals.

The loading for an internal bay is as follows:

Roof – offices and showroom: dead load $= 6.0 \text{ kN/m}^2$

imposed load $= 1.5 \text{ kN/m}^2$

Floors – offices   dead load
slab, finish, ceiling, services, partitions $= 7.0 \text{ kN/m}^2$

Floors – offices: imposed load $= 2.5 \text{ kN/m}^2$

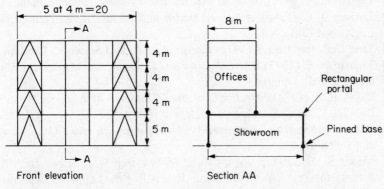

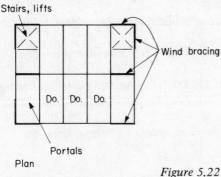

Figure 5.22

External walls at roof level   = 2.6 kN/m

External walls at floor levels   = 8.0 kN/m

Wind load – CP3, Chapter V, Part 2 [14].

Design work required:

(a) Estimate the loading on an internal portal in A–A.
(b) Analyse and design the portal.
(c) Make a detail drawing of the portal.

References and further reading

[1] Bresler, B and Lin, T. Y. (1964). *Design of Steel Structures*. John Wiley, New York.

[2] Coates, R. C., Coutie, M. G. and Kong, F. K. (1972). *Structural Analyses*. Nelson, London.

[3] Ghali, A. and Neville, A. M. (1978). *Structural Analysis*. Chapman and Hall, London.

[4] *I.C.L. Computer Manuals* (1969). Analysis of Plane Frames and Grids TP4178. Analysis of Space Frames TP4220. International Computers Ltd.

[5] Lightfoot, E. (1961). *Moment Distribution*. E. & F. N. Spon, London.

[6] Lothers, J. E. (1960). *Advanced Design in Structural Steel*. Prentice Hall, Englewood Cliffs, N.J.

[7] *Steel Designers Manual* (1972). Constrado, Crosby Lockwood, London.

[8] Trahair, N. S. (1977). *The Behaviour and Design of Steel Structures*. Chapman and Hall, London.

[9] Wood, R. H. (1974). *Effective Lengths of Columns in Multi-storey Buildings*. Building Research Establishment, Department of the Environment, London. Printed in *Structural Engineer* (1974) **52**, nos. 7, 8 and 9 (July, August and September).

[10] Wood, R. H. (1974). *Rigid-jointed Multi-storey Steel Frame Design: A State-of-the-Art Report*. Building Research Establishment, Department of the Environment, London.

[11] BS449, Part 2 (1969). *The use of structural steel in building*. British Standards Institution.

[12] Sherbourne, A. N. (1961). Bolted beam to column connections. *The Structural Engineer* **39** (6).

[13] *Structural steelwork handbook* (1978). B.C.S.A. and Constrado, London.

[14] CP3, Chapter V, Part 2 (1972). *Loading wind loads*. British Standards Institution, London.

[15] Horne, M. R. (1971). *Plastic theory of structures*. Nelson, London.

6 Rigid elastic design – examples

6.1 Introduction

Two examples of rigid structures designed in accordance with elastic theory to BS449 [1] are presented. These are:

(a) a two-storey school building;
(b) a single-storey industrial building.

The two-storey building consists of rigid frames transversely and longitudinally to resist wind load. This eliminates bracing altogether. Other construction is in simple design. The analysis using a computer program and design is given for a transverse portal. The main constructional details are shown on sketches.

The single-storey industrial building includes a 600 kN electric overhead travelling crane. The building consists of non-uniform rigid fixed-base portals transversely. Bracing is provided in the longitudinal direction to give stability. The analysis by computer program and design is given for a transverse portal. The main frame details are shown on sketches.

6.2 Elastic design — two-storey building

6.2.1 Specification

An infants playschool is to be in a two-storey building with the upper storey generally in open plan with subdivision in lightweight partitions. All walls except stairs and toilet areas are in double glazing to give maximum natural lighting. *In situ* slabs will be used for roof and floor with suspended ceilings under. The staircase walls and toilet walls will be in breeze block. Fire protection is in lightweight casing for the columns.

The framing plans are shown in Fig. 6.1. Rigid frames are provided in each direction to resist wind loads. This gives wall panels clear of bracing all round. The other steel beams are simply connected.

The following work is to be carried out:

(a) estimate the loading on the internal transverse rigid frame due to dead, imposed and wind loads;

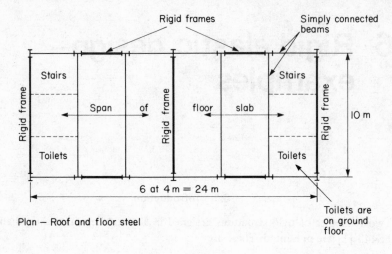

Plan — Roof and floor steel

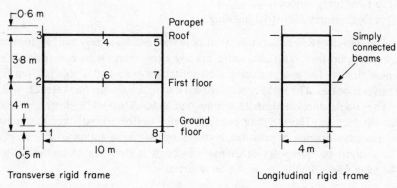

Figure 6.1

(b) analyse the rigid frame for the separate load cases using the I.C.L. plane frame program. Draw the bending moment diagrams for each load case;
(c) design the transverse rigid frame;
(d) show the structural steel details on sketches.

The material to be used is Grade 43 Steel. All rigid frame joints are to be in HSFG bolts.

6.2.2 Loading

(a) Dead and imposed loads

(i) Roof (kN/m^2)

    Asphalt and chippings   0.5

    Screed   0.5

Slab 125 mm thick	2.95
Steel	0.4
Fireproof ceiling	0.35
Total dead load	4.7
Imposed load	1.5
Total roof load	6.2

		(kN/m)
Roof beam: Dead	= 4.7 x 4	18.8
Imposed	= 1.5 x 4	6.0
Total		24.8

(ii) First floor

	(kN/m²)
Floor tiles	0.15
Screed	0.6
Slab 175 mm thick	4.13
Steel	0.5
Fire proof ceiling	0.4
Services	0.1
Lightweight partitions	1.0
Total dead load	6.88
Imposed load	3.0
Total roof load	9.88

		(kN/m)
Floor beam: Dead	= 6.88 x 4	27.52
Imposed	= 3 x 4	12.00
Total		39.52

The side wall construction is shown in Fig. 6.2. The estimated loading for

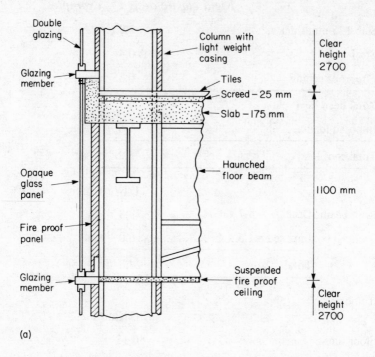

(a)

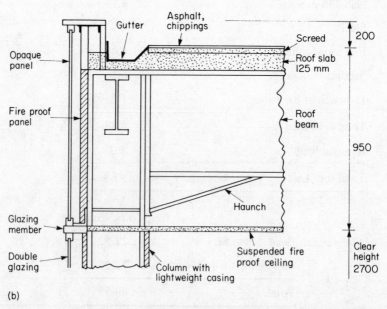

(b)

Figure 6.2 (a) Section through floor at the transverse rigid frame. (b) Section at roof level at the transverse rigid frame.

the side wall is:

Double glazing	$= 0.6 \text{ kN/m}^2$
Opaque panel fire proof panel	$= 0.6 \text{ kN/m}^2$
Curtain wall supports	$= 0.2 \text{ kN/m}$
Side beam	$= 0.5 \text{ kN/m}$
Parapet support steel	$= 0.3 \text{ kN/m}$
Wall panel on the inside of the parapet	$= 0.3 \text{ kN/m}^2$

Column loads

Roof level – parapet + side cladding + support steel + side beam + 250 mm strip of roof slab and ceiling
$$= (0.3 \times 9.8) + (4 \times 0.5) + (1.15 \times 4 \times 0.6) + (1.05 \times 0.3) + (8 \times 0.2)$$
$$+ (0.25 \times 4 \times 4.3) = 13.8 \text{ kN.}$$

First floor level – glazing + side cladding + support steel + side beam + 250 mm of floor slab and ceiling
$$= (0.6 \times 4 \times 2.7) + (0.6 \times 4 \times 1.1) + (8 \times 0.2) + (4 \times 0.5) + (0.25 \times 5.25 \times 4)$$
$$= 16.9 \text{ kN.}$$

Column self weight + lightweight casing	$= 1.4 \text{ kN/m}$
Roof to first floor $= 3.8 \times 1.4$	$= 5.3 \text{ kN}$
First floor to base $= 4 \times 1.4$	$= 5.6 \text{ kN}$

The load diagrams for the internal frame are shown in Fig. 6.3(a) and (b) for dead and imposed load, respectively.

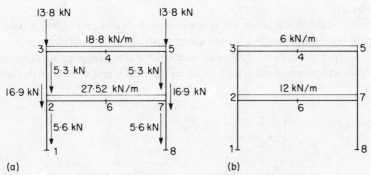

Figure 6.3   (a) Dead load. (b) Imposed load.

(b) Wind loads   See CP3, Chapter V, Part 2 [2].

Location – North East England

Basic wind speed $= 45 \text{ m/s}$

162   Steel structures

Topography factor $S_1 = 1.0$

The building is situated on the outskirts of a city with obstructions up to 10 m high.

Ground roughness – Category 3

Building size – Class B (both the maximum horizontal and vertical dimensions do not exceed 50 m)

Statistical factor $S_3 = 1.0$

The heights for roof and first floor, the factor S_2 from Table 3 of CP3, the design wind speeds and dynamic pressures are given in Table 6.1

Table 6.1   Wind pressures

Location	Height	S_2	Design wind speed V_s(m/s)	Dynamic pressure (kN/m²)
Roof	7.9 m	0.702	0.702 x 45 = 31.6	0.613 x 31.6²/10³ = 0.61
First floor	3.5 m	0.613	0.613 x 45 = 27.6	0.613 x 27.6²/10³ = 0.47

Pressure coefficients
The external pressure coefficients from Tables 7 and 8, CP3, Chapter V, Part 2, are shown in Fig. 6.4. The internal pressure coefficients are taken from Appendix E, CP3, for the case where there is a negligible probability of a dominant opening occurring during a severe storm. C_{pi} is then taken as the more onerous of the values −0.2 or −0.3.

Wind loads
The wind loads on the walls are applied to the frame at roof and first floor level through the curtain walling which spans vertically from roof to first floor and first floor to ground floor. The wind on the parapet introduces a small moment at roof level which is neglected. The uplift on the roof is taken as a distributed load.

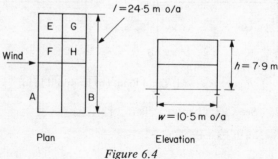

Figure 6.4

The frame, its dimensions for calculating loads, the pressure coefficients and dynamic pressures are shown in Fig. 6.5(a).

Using this information, the wind loads on the frame for the two cases of internal pressure and internal section are calculated. The frame carries the horizontal wind loading on one half the length of the building, i.e. 12 m length. The vertical uplift on the frame is from a 4 m length of roof.

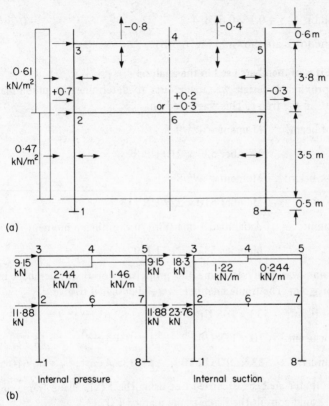

(a)

(b)

Figure 6.5 (a) Dynamic pressure, pressure coefficients and dimensions for calculating wind loads. (b) Wind loads.

Internal pressure case:

Load at 3 = 0.61 × 12 × 2.5 × 0.5	= 9.15 kN
Load at 5 = 0.61 × 12 × 2.5 × 0.5	= 9.15 kN
Load at 2 = 0.5 × 12[1.9 × 0.61 + 1.75 × 0.47]	= 11.88 kN
Load at 7	= 11.88 kN
Load on 3–4 = 0.61 × 4 × 1.0	= 2.44 kN/m
Load on 4–5 = 0.61 × 4 × 0.6	= 1.46 kN/m.

Internal suction case:

Load at 3 = 0.61 x 12 x 2.5	= 18.3 kN
Load at 2 = 2 x 11.88	= 23.76 kN
Loads at 5 and 7 are zero	
Load on 3–4 = 0.5 x 0.61 x 4	= 1.22 kN/m
Load on 4–5 = 0.1 x 0.61 x 4	= 0.244 kN/m.

The wind loads are shown in Fig. 6.5(b).

6.2.3 Sizes of members used in the analysis

An approximate design was made first to determine the preliminary sizes of members in the frame. This was based on:

Roof beam: Moment = $wl^2/9$

Member 533 x 210 UB 89,

Floor beam: Moment = $wl^2/9$

Member 610 x 229 UB 113,

Column: Axial load + one third of floor beam moment

Member 533 x 210 UB 92.

The frame was analysed using the properties for these members in the computer program. The frame members were re-designed to give:

Roof beam: 533 x 165 UB 66, $A = 83.6 \text{ cm}^3$; $I = 35\,083 \text{ cm}^4$,

Floor beam: 610 x 178 UB 82, $A = 104.4 \text{ cm}^2$; $I = 55\,779 \text{ cm}^4$,

Columns: 533 x 210 UB 109, $A = 138.4 \text{ cm}^2$; $I = 66\,610 \text{ cm}^4$.

The computer analysis is carried out using the I.C.L. Plane frame program. The reader should consult the Plane frame manual [3].

6.2.4 Computer analysis

The computer analysis is carried out using the I.C.L. Plane frame program. The reader should consult the Plane frame manual [3].

(a) Input data

The joint numbering is shown in Fig. 6.6. The sign convention for forces and moments is also shown in the figure. The input data is given in Table 6.2.

(b) Output

The output for the five load cases is given in Table 6.3. The bending moment

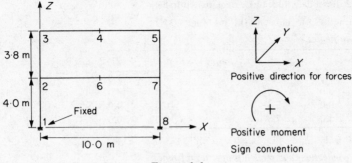

Figure 6.6

diagrams for these load cases are shown in Fig. 6.7. The load combinations required can then be seen clearly.

6.2.5 Design of the frame members

(a) Roof beam 3-4-5

    Joint 3 — Dead + imposed loads on roof and floor:

        Moment $M = 144.8 + 44.5 + 2.4$ $= 191.7$ kN m

        Shear V $= 94.0 + 30.0$ $= 124.0$ kN

        Thrust P $= 71.9 + 14.6 + 11.4$ $= 97.9$ kN.

Support the bottom flange at 2.5 m from 3 by stays from the floor slab. Try 533 x 165 UB 66.

    $A = 83.6$ cm^2, $r_{YY} = 3.21$ cm

    $Z_{XX} = 1337$ cm^2, $D/T = 45.6$

    $l/r_{YY} = 2500/32.1 = 78.$

Allowable stresses:

        $p_c = 106$ N/mm^2

        $p_{bc} = 165$ N/mm^2

Axial stress:

        $f_c = 97.9 \times 10/83.6$ $= 11.7$ N/mm^2.

Bending stress:

        $f_{bc} = 191.7 \times 10^3/1337$ $= 143.4$ N/mm^2.

Combined:

        $(11.7/106) + (143.4/165) = 0.11 + 0.87$ $= 0.98.$

166   Steel structures

Table 6.2 Input data. Fixed-base rigid-jointed frame

Units   lengths (m)   moments (kN m)   forces (kN)

Joint co-ordinates

Joint	X	Z	Joint	X	Z
1	0.0	0.0	5	10.0	7.8
2	0.0	4.0	6	5.0	4.0
3	0.0	7.8	7	10.0	4.0
4	5.0	7.8	8	10.0	0.0

Option   rigid

Member connections and section properties reference

End joints				Section properties reference
1–2	2–3	5–7	7–8	A
3–4		4–5		B
2–6		6–7		C

Young's modulus 21 000.0 kN/cm^2

Properties reference	Area (cm^2)	Moments of inertia (cm^4)
A	138.4	66 610.0
B	83.6	35 083.0
C	104.8	55 779.0

Restraints   joints 1, 8 ALL (fixed)

Load data – lengths

Load case	Joint A	Joint B	Type	Start value	Direction axis	Start position	End value	End position
Dead	3	4	LDA	18.8	−Z	0.0	18.8	5.0
	4	5	LDA	18.8	−Z	0.0	18.8	5.0
	2	6	LDA	27.5	−Z	0.0	27.5	5.0
	6	7	LDA	27.5	−Z	0.0	27.5	5.0
	2		LP	22.2	−Z			
	7		LP	22.2	−Z			
	3		LP	13.8	−Z			
	5		LP	13.8	−Z			
Imposed, roof	3	4	LDA	6.0	−Z	0.0	6.0	5.0
	4	5	LDA	6.0	−Z	0.0	6.0	5.0
Imposed, floor	2	6	LDA	12.0	−Z	0.0	12.0	5.0
	6	7	LDA	12.0	−Z	0.0	12.0	5.0
Wind, internal pressure	2		LP	11.88	+X			
	3		LP	9.15	+X			
	5		LP	9.15	+X			
	7		LP	11.88	+X			
	3	4	LDA	2.44	+Z	0.0	2.44	5.0
	4	5	LDA	1.46	+Z	0.0	1.46	5.0
Wind, internal suction	2		LP	23.76	+X			
	3		LP	18.3	+X			
	3	4	LDA	1.22	+Z	0.0	1.22	5.0
	4	5	LDA	0.244	+Z	0.0	0.244	5.0

Table 6.3   Computer output

Load case	Joints A−B	Joint A			Joint B		
		Thrust	Shear	Moment	Thrust	Shear	Moment
Dead	1−2	267.5	− 33.1	45.7	−267.5	33.1	86.7
	2−3	107.8	− 71.9	128.7	−107.8	71.9	144.8
	3−4	71.9	94.0	−144.8	− 71.9	0.0	− 90.2
	2−6	− 38.9	137.5	−215.4	38.9	0.0	−128.3
Imposed, roof	1−2	30.0	3.2	− 3.5	− 30.0	− 3.1	− 9.1
	2−3	30.0	− 14.6	11.0	− 30.0	14.6	44.5
	3−4	14.6	30.0	− 44.5	− 14.6	0.0	− 30.5
	2−6	− 17.7	0.0	− 1.9	17.7	0.0	1.9
Imposed, floor	1−2	60.0	− 18.8	24.7	− 60.0	18.8	50.3
	2−3	0.0	− 11.4	41.1	0.0	11.4	2.4
	3−4	11.4	0.0	− 2.4	− 11.4	0.0	2.4
	2−6	7.3	60.0	− 91.4	− 7.3	0.0	− 58.6
Wind, internal pressure	1−2	− 22.6	20.0	− 59.6	22.6	− 20.0	− 20.4
	2−3	− 15.1	13.1	− 17.5	15.1	− 13.9	− 35.3
	3−4	− 4.7	− 15.1	35.3	4.7	2.9	9.9
	4−5	− 4.7	− 2.9	− 9.9	4.7	− 4.4	6.4
	2−6	5.8	− 7.5	37.9	− 5.8	− 7.5	− 0.6
	6−7	5.8	− 8.9	0.6	− 5.8	7.5	36.7
	5−7	− 4.4	4.4	− 6.4	4.3	− 4.4	− 10.4
	7−8	3.1	22.1	− 26.3	− 3.4	− 22.1	− 61.9
Wind, internal suction	1−2	− 16.5	21.1	− 61.3	16.5	− 21.1	− 22.9
	2−3	− 9.0	10.8	− 14.8	9.0	− 10.8	− 26.2
	3−4	7.5	− 9.0	26.2	− 7.5	2.9	3.8
	4−5	7.5	− 2.9	− 3.8	− 7.5	1.7	15.4
	2−6	13.5	− 7.5	37.7	− 13.5	7.5	− 0.3
	6−7	13.5	− 7.5	0.3	− 13.5	7.5	36.9
	5−7	1.7	7.5	− 15.4	− 1.7	− 7.5	− 13.2
	7−8	9.2	20.9	− 23.8	− 9.2	− 20.9	− 60.1

	Deflections					
Load	Direction	Joint	Value	Joint	Value	
Dead	Z	4	−0.962 cm	6	−0.835 cm	
Imposed, roof	Z	4	−0.329 cm	6	+0.017 cm	
Imposed, floor	Z	4	+0.034 cm	6	−0.385 cm	
Wind, internal pressure	X	3	+0.423 cm	5	+0.424 cm	
Wind, internal suction	X	3	+0.423 cm	5	+0.425 cm	

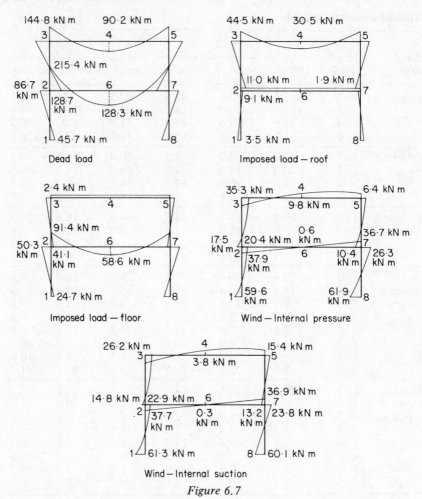

Figure 6.7

Shear stress:

$$f_s = 124 \times 10^3 / (524.8 \times 8.8) \qquad\qquad = 26.9 \text{ N/mm}^2.$$

Note that a small haunch is provided at the end of the beam to form the joint. This will reduce stresses here.

Deflection at joint 4

$$\delta = 9.62 + 3.29 \qquad\qquad = 12.91 \text{ mm}$$

$$\delta/\text{span} \qquad\qquad = 12.91/10\,000 \qquad\qquad = 1/774$$

Use 533 × 165 UB 66.

(b) Floor beam 2-6-7

Joint 2 — Dead + imposed load on floor only:

Moment M = 215.4 + 91.4		= 306.8 kN m
Shear V = 137.5 + 60.0		= 197.5 kN
Tension T = 38.9 + 7.3		= 31.6 kN.

Support the bottom flange at 2.5 m from 2 by stays from the floor slab. Try 533 x 210 UB 92.

$$A = 117.8 \text{ cm}^2, \qquad r_{YY} = 4.38 \text{ cm}$$

$$Z_{XX} = 2076 \text{ cm}^3, \qquad D/T = 34.1$$

$$l/r_{YY} = 2500/43.8 = 57.1.$$

Allowable stresses:

$$p_t = 155 \text{ N/mm}^2$$

$$p_{bc} = p_{bt} = 165 \text{ N/mm}^2.$$

Axial stress:

$$f_t = 31.6 \times 10/117.6 \qquad\qquad = 2.7 \text{ N/mm}^2.$$

Bending stress:

$$f_{bt} = 306.8 \times 10^3/2072 \qquad\qquad = 148.1 \text{ N/mm}^2.$$

Combined:

$$(2.7/155) + (148.1/165) = 0.02 + 0.9 \qquad = 0.92.$$

Shear stress:

$$f_s = 197.5 \times 10^3/(533.1 \times 10.2) \qquad\qquad = 36.3 \text{ N/mm}^2.$$

Note that the beam size used in the analysis, 610 x 178 UB 82, is overstressed. The moment of inertia has changed from 55 779 cm^4 to 55 225 cm^4.

Deflection at joint 6

δ = 8.35 + 3.85	= 12.2 mm.
δ/span = 12.2/10 000	= 1/819

Use 533 x 210 UB 92.

(c) Column 1-2-3

The column will be made uniform throughout.

Joint 2–1 – Dead + imposed load on the floor:

Moment M	= 86.7 + 50.3	= 137.0 kN m
Thrust P	= 267.5 + 60.0	= 327.5 kN
Shear V	= 33.1 + 18.8	= 51.9 kN.

The maximum wind moment here is 26.3 kN m or 19.2% of the moment due to dead and imposed loads.

Joint 2–3 – Dead + imposed load on the roof and floor:

Moment M	= 128.7 + 11 + 41.1	= 180.8 kN
Thrust P	= 107.8 + 30.0	= 137.8 kN
Shear V	= 71.9 + 14.6 + 11.4	= 97.9 kN.

Joint 3 Dead + imposed load on the roof and floor

Moment M = 144.8 + 44.5 + 2.4		= 191.7 kN m
Thrust P = 107.8 + 30.0		= 137.8 kN
Shear V = 71.9 + 14.6 + 11.4		= 97.9 kN

Try 533 x 210 UB 92

A = 117.8 cm^2	r_{XX} = 21.68 cm	D/T = 34.1
Z_{XX} = 2076 cm^3	r_{YY} = 4.51 cm	I_{XX} = 55 353 cm^4

The effective lengths for the columns for axial load are estimated for both directions.

(i) Transverse direction, i.e. in the plane of the frame. The effective lengths are calculated using Dr Wood's chart which was reproduced in Fig. 5.5(b). The stiffnesses and effective lengths are shown on Fig. 6.8(a).

(ii) Longitudinal direction, i.e. at right angles to the plane of the frame. The stability is provided by rigid frames (refer to Fig. 6.1). The effective length of the columns in the frame under consideration, laterally, will be taken to be the same as that of the side rigid frames. Assume that a constant section throughout is used for the side frames. The stiffnesses and effective lengths are shown on Fig. 6.8(b). The column lengths are checked separately.

1. Portion 1–2 Joint 2

l/r_{XX} = 6400/213.2		= 30
l/r_{YY} = 5440/45.1		= 121
p_c = 59 N/mm^2		

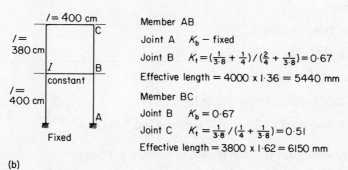

$I = 35\,083 \text{ cm}^4$
$l = 1000 \text{ cm}$ $K = 35 \cdot 1$

$I = 47\,491 \text{ cm}^4$

$K = 125$

$l = 380 \text{ cm}$

$I = 55\,225 \text{ cm}^4$
$l = 1000 \text{ cm}$ $K = 55 \cdot 2$

$I = 47\,491 \text{ cm}^4$

$K = 118 \cdot 7$

$l = 400 \text{ cm}$

1 ⊿ Fixed

Member 1–2   Joint 1 K_b – fixed
              Joint 2 $K_t = (125 + 118 \cdot 7)/(125 + 118 \cdot 7 + 55 \cdot 2) = 0 \cdot 82$
              Effective length $= 1 \cdot 6 \times 4000 = 6400$ mm

Member 2–3   Joint 2 $K_b = 0 \cdot 82$
              Joint 3 $K_t = 125/(125 + 35 \cdot 1) = 0 \cdot 78$
              Effective length $= 2 \cdot 5 \times 3800 = 9500$ mm

(a)

$l = 400 \text{ cm}$

$l = 380 \text{ cm}$

I constant

$l = 400 \text{ cm}$

Fixed

Member AB
Joint A K_b – fixed
Joint B $K_t = (\frac{1}{3 \cdot 8} + \frac{1}{4})/(\frac{2}{4} + \frac{1}{3 \cdot 8}) = 0 \cdot 67$
Effective length $= 4000 \times 1 \cdot 36 = 5440$ mm
Member BC
Joint B $K_b = 0 \cdot 67$
Joint C $K_t = \frac{1}{3 \cdot 8}/(\frac{1}{4} + \frac{1}{3 \cdot 8}) = 0 \cdot 51$
Effective length $= 3800 \times 1 \cdot 62 = 6150$ mm

(b)

Figure 6.8   (a) Effective lengths for frame in the transverse direction. (b) Effective lengths for the frame in the longitudinal direction.

For the allowable bending stress

$l/r_{YY} = 4000/45.1 = 88.5;\ D/T = 34.1$

$p_{bc} = 165 \text{ N/mm}^2$

$f_c = 327.5 \times 10/117.8$ $= 27.8 \text{ N/mm}^2$

$f_{bc} = 137.10^3/2076$ $= 66.3 \text{ N/mm}^2$

Combined $(27.8/59) + (66.3/165)$ $= 0.88$

2. Portion 2–3 Joint 3

$l/r_{XX} = 9500/213.2$ $= 44.5$

$l/r_{YY} = 6150/45.1$ $= 137$

$p_c = 48 \text{ N/mm}^2$

For the allowable bending stress

$l/r_{YY} = 3800/45.1 = 84.5; D/T = 34.1$

$p_{bc} = 165 \text{ N/mm}^2$

$f_c = 137.5 \times 10/117.8$ $= 11.7 \text{ N/mm}^2$

$f_{bc} = 191.7 \times 10^3/2076$ $= 92.5 \text{ N/mm}^2$

Combined $(11.7/48) + (92.5/165)$ $= 0.81$

The column section chosen is satisfactory. Note that the column size has been reduced from 533 x 210 UB 109 used in the analysis.

Column section 533 x 210 UB 92.

(d) Joints

1. Floor beam to column

Moment M = 306.8 kN m

Shear V = 197.5 kN

The axial load has been neglected.

The arrangement of the joint is shown in Fig. 6.9(a).

Moment. Assume that this is resisted by the top 4 bolts

$T = 306.8/0.9 \times 4$ $= 85.2 \text{ kN}$

Try M20, HSFG Bolts − proof load $= 144 \text{ kN}$

Permissible tension $= 0.6 \times 144$ $= 86.4 \text{ kN}$

Shear. Provide 6 No. M20, HSFG bolts at 46.3 kN per bolt

Shear resistance $= 6 \times 46.3$ $= 277.8 \text{ kN}$

2. Roof beam to column

Moment M = 191.7 kN m

Shear V = 124   kN

The arrangement of the joint is shown in Fig. 6.9(b).

Moment. Assume this is resisted by the top four bolts:

$T = 191.7/4 \times 0.6$ $= 79.8 \text{ kN.}$

Provide four no. M20 HSFG bolts.

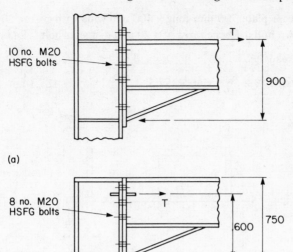

(a)

(b)

Figure 6.9   (a) Floor beam to column joint. (b) Roof beam to column joint.

Shear. Provide four no. M20 HSFG bolts:

Shear resistance = 4 x 46.3   = 185.2 kN.

(e) Column base plate
    Case I – Dead + imposed loads:

   $M = 45.7 + 3.5 + 24.7$ = 66.9 kN m

   $P = 267.5 + 30 + 60$ = 357.5 kN m.

Case II – Dead + wind internal pressure:

   $M = 45.7 + 61.9$ = 107.6 kN m

   $P = 267.5 + 3.1$ = 270.6 kN.

The allowable stresses are:

Concrete, Grade 25
    Allowable bearing stress p_b = 4.7 N/mm^2

Bolts, Grade 4.6
    Allowable tensile stress p_t = 120 N/mm^2.

The allowable stresses including 25% increase for wind are

Concrete p_b =   5.87 N/mm^2

Bolts p_t = 150 N/mm^2.

Assume a base plate 700 mm long x 400 mm wide with six no. 20 mm diameter holding down bolts. The net area of a 20 mm diameter bolt is 245 mm^2.

(i) Case I loads

Eccentricity e = 66.9 x 10^3/357.5 = 187.1 mm

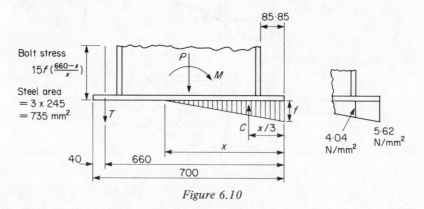

Figure 6.10

Tensions occurs in the bolts. The loading and stress distribution are shown in Fig. 6.10.

$$P = \tfrac{1}{2}fx \times 400 - 15f\left(\frac{660 - x}{x}\right)735$$

$$= 200fx - 7.276 \times 10^6 f/x - 11025$$

$$M = 187.1P$$

$$= 37\,420fx - 1.361 \times 10^9 f/x - 2.063 \times 10^6 \qquad (6.1)$$

$$= 200fx\left[350 - \frac{x}{3}\right] + 7.276 \times 10^6 f/x \times 310 - 11\,025f \times 310$$

$$= 70\,000fx - 66.7fx^2 + 2.255 \times 10^9 f/x - 3.418 \times 10^6 f \qquad (6.2)$$

Equate (6.1) and (6.2) and reduce to give:

$$x^3 - 488.5x^2 - 82\,174x - 5.421 \times 10^7 = 0.$$

Solving gives x = 530 mm.

Concrete stress f = 3.46 N/mm^2.

(ii) Case II loads

Eccentricity e = 107.6 x 10^3/270.6 = 396.6 mm

$$M = 397.6P$$

$$= 79\,520fx - 2.893 \times 10^9 f/x - 4.384 \times 10^6 \qquad (6.3)$$

$$= 70\,000fx - 66.7fx^2 - 2.255 \times 10^9 f/x - 3.418 \times 10^6 f \qquad (6.4)$$

$$x^3 - 142.7x^2 - 116\,972x - 7.718 \times 10^7 \qquad = 0$$

Solving gives x = 305 mm.

Concrete stress f = 5.62 N/mm^2.

The size of base is satisfactory.

Stress in the bolts = 15 x 5.62 x 355/305   = 98.1 N/mm^2. Safe.

Design the base to cantilever from the column.

M = 5.62 x 85.85^2/3 + 4.04 x 85.85^2/6   = 1.88 x 10^4 = N mm.

Allowable stress in slab base = 165 + 25%   = 207 N/mm^2.

Slab thickness t = (6 x 1.88 x 10^4/207)$^{1/2}$ = 23.4 mm.

Provide a base slab 25 mm thick.

The weld between column and slab — the maximum load per mm length — is given by

107. 6 x 10^6/(208.7 x 515.1) − 270.6 x 10^3/(2 x 208.7) = 352.6 N/mm^2.

Provide 8 mm fillet weld on flanges and 5 mm fillet weld on both sides of the web.

The column base is shown in Fig. 6.11

(e) Frame arrangement
The frame arrangement is shown in Fig. 6.11.

6.3 Elastic design — rigid frame factory building

6.3.1 Specification
The section through a rigid frame factory building is shown in Fig. 6.12. The frames are at 10 m centres and the length of the building is 60 m. Preliminary dimensions of members are also shown. The building loads are:

Roof — dead load measure on slope:

  sheeting, insulation, purlins   = 0.57 kN/m^2

  portal rafter   = 1.8 kN/m

Roof — imposed load on plan: = 0.75 kN/m^2

Walls — cladding, insulation, sheeting rails   = 0.6 kN/m^2

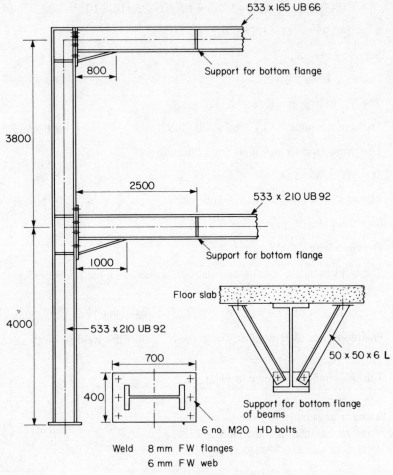

Weld   8 mm F W flanges
        6 mm F W web
Base plate — 700 x 400 x 25 plate

Figure 6.11

Column — roof leg	= 8 kN
crane portion	= 40 kN
Crane girder, surge girder, rail, etc.	= 50 kN

The wind loading is to be in accordance with CP3, Chapter V, Part 2 [2].
The location of the building is in the North East of England on the outskirts of
a city:

The crane data are:

Capacity	= 600 kN

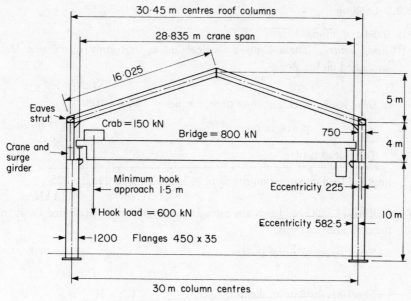

Figure 6.12

Span	= 28.835 m
Weight of the crane bridge	= 800 kN
Weight of the crab	= 150 kN
End carriage wheel centres	= 3.5 m
Minimum hook approach	= 1.5 m
End clearance	= 400 m
Number of wheels in the end carriage	= 2

The following work is to be carried out:

(a) Estimate the loading on an internal portal due to dead load, imposed load, crane wheel loads, crane surge, wind internal pressure and wind internal suction.
(b) Analyse the frame for the six separate load cases listed in (a) using the I.C.L. plane frame program and draw the bending moment diagram for each case.
(c) Design the frame members:
    (i) crane column and base;
    (ii) roof leg;
    (iii) rafter.
(d) Show the main constructional details on sketches.

6.3.2 Loading

(a) Dead and imposed loads

(i) *Roof loads.*   These loads are calculated as uniformly distributed loads measured on the slope.

		(kN/m)
Dead load – sheeting, insulation, purlins	0.57 x 10 = 5.7	
portal rafter	1.8	
Total dead load	7.5	

Imposed load measured on the slope = 0.75 x 10 x 15 225/16 025

$$= 7.13 \text{ kN/m.}$$

(ii) *Wall loads.*   These loads are calculated as uniformly distributed loads for the computer analysis.

Eaves to crane girder level:

	(kN/m)
Sheeting, insulation, sheeting rails	0.6 x 10 = 6
Roof leg	4 kN = 2
Total wall load	8

Crane girder level to the base:

	(kN/m)
Sheeting, insulation, sheeting rails	0.6 x 10 = 6
Crane column	40 kN = 4
Total wall load	10

(iii) Crane girder including surge girder rail, etc.   = 50 kN

Moment due to eccentricity = 50 x 0.58   = 29 kN m

The dead and imposed loads are shown in Fig. 6.15(a) and (b), respectively.

(b) Crane loads

(i) *Vertical crane wheel loads.*

See Fig. 6.13(a). The crane wheel loads including 25% for impact are:

Heavy side = 0.5 x 1.25 [400 + 750 x 27.3/28.8]   = 694 kN

Light side = 0.5 x 1.25 [400 + 750 x 1.5/28.8]   = 275 kN.

The end carriage wheel loads are shown in Fig. 6.13(b).

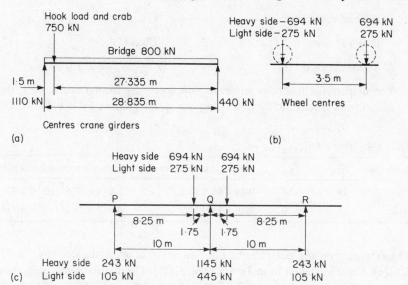

Figure 6.13 (a) Static crane wheel loads. (b) End carriage wheel loads. (c) Column reactions.

See Fig. 6.13(c). The column reactions are:

Heavy side R_Q = 2 x 694 x 8.25/10   = 1145 kN

Light side R_Q = 2 x 275 x 8.25/10   =   445 kN

The moments due to the crane loads are:

Heavy side = 1145 x 0.58   = 665 kN m

Light side = 445 x 0.58   = 258 kN m

The vertical crane wheel loads and moments are shown in Fig. 6.15(c).

(ii) *Horizontal crane loads* — 10% of crab + hook load.

Horizontal wheel load = 0.1 (150 + 600)/4   = 18.75 kN

The load has been divided equally between the four wheels. See Fig. 6.13(c). The maximum column reaction is:

R_Q = 2 x 18.75 x 8.25/10   = 31 kN.

The column reaction is the same on each side of the frame. The horizontal crane loads are shown in Fig. 6.15(d).

(c) Wind loads (See CP3, Chapter V, Part 2[2]).

Basic wind speed V = 45 m/s

Topography factor S_1 = 1.0

Ground roughness   Category 3

Building size   Class C

Statistical factor S_3 = 1.0

The heights for the ridge and eaves, the factor S_2 from Table 3, CP3, the design wind speeds and dynamic pressures, q, are given in Table 6.4.

Table 6.4 Wind pressures

Location	Height	S_2	Design wind speed (m/s)	Dynamic pressure, q (kN/m^2)
Roof	19 m	0.84	0.84 x 45 = 37.8	$0.613 \times 37.8^2/10^3 = 0.876$
Walls	14 m	0.76	0.76 x 45 = 34.2	$0.613 \times 34.2^2/10^3 = 0.718$

The external pressure coefficients, C_{pe}, for the roof and walls are shown in Fig. 6.14(a). These are taken from Tables 7 and 8, CP3. The internal wind coefficients, C_{pi}, are taken from Appendix E, CP3, for the case where there is a negligible probability of a dominant opening occurring during a severe storm. C_{pi} is taken as the more onerous of the values −0.2 or −0.3. The total pressure coefficients are shown in Fig. 6.14(b).

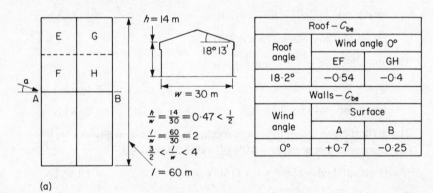

(a)

Roof – C_{be}		
Roof angle	Wind angle 0°	
	EF	GH
18·2°	−0·54	−0·4

Walls – C_{be}		
Wind angle	Surface	
	A	B
0°	+0·7	−0·25

$h = 14$ m

$18° 13'$

$w = 30$ m

$\frac{h}{w} = \frac{14}{30} = 0.47 < \frac{1}{2}$

$\frac{l}{w} = \frac{60}{30} = 2$

$\frac{3}{2} < \frac{l}{w} < 4$

$l = 60$ m

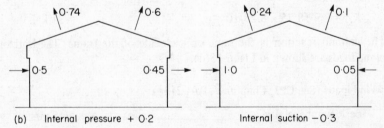

(b)   Internal pressure + 0·2   Internal suction −0·3

Figure 6.14   (a) External pressure coefficients. (b) Total pressure coefficients.

The wind load on the roof slopes and walls is given by the expression

$$10q(C_{pe} - C_{pi}) \qquad \text{kN/m.}$$

The wind loads are shown in Fig. 6.15(e) and (f) for the internal pressure and internal suction cases, respectively.

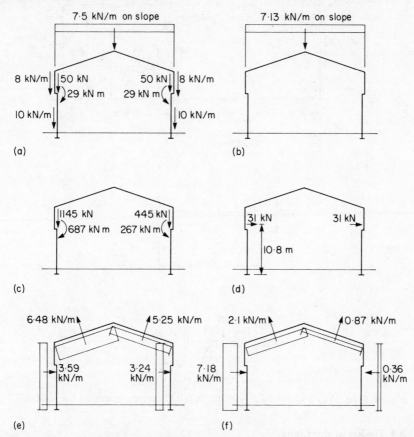

Figure 6.15 (a) Dead loads. (b) Imposed loads. (c) Vertical crane wheel loads. (d) Crane surge loads. (e) Wind load internal pressure. (f) Wind load internal suction.

6.3.3 Computer analysis
The frame dimensions for the computer analysis and the assumed sections and properties are shown in Fig. 6.16. These sections have been arrived at by carrying out several analyses and designs. The frame loading is taken from Fig. 6.15.

The frame is analysed for the separate load cases using the I.C.L. Plane frame progam [3]. The bending moment diagrams from the analysis are shown in Fig. 6.17. The axial loads required for design are given in the design of the separate elements.

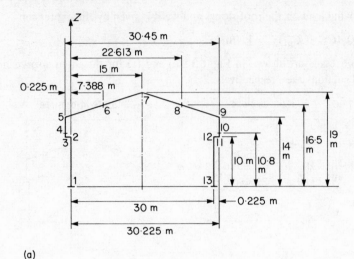

(a)

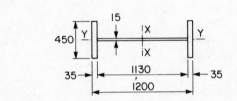

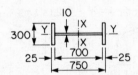

Figure 6.16   (a) Joints and co-ordinates. (b) Sections assumed for analysis.

6.3.4 Design of the frame

(a) Crane column 1-2

The design actions are:

Base — dead + imposed + crane wheels + crane surge:

$P = 302.2 + 114.2 + 1142.6 + 2.0$ = 1561 kN

$M = 663.7 + 632.3 + 205.0 - 304.3$ = 1395.3 kN m.

Crane girder — dead + imposed + crane wheels:

$P = 202.5 + 114.2 + 1142.6 + 2.0$ = 1461.3 kN

$M = 86.6 + 69.0 + 481.8 + 5.7$ = 643.1 kN m.

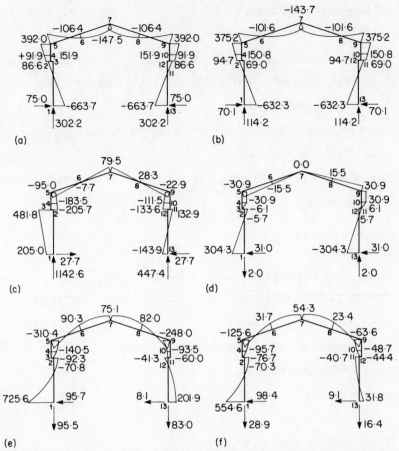

Figure 6.17 (a) Dead loads. (b) Imposed loads. (c) Vertical crane wheel loads. (d) Crane surge loads. (e) Wind – internal pressure. (f) Wind – internal suction. Moments in kN m, reactions in kN.

The trial section and properties are shown in Fig. 6.16(b). For dealing with a built-up column and the effective lengths for a crane column see Section 3.4.2.

$l/r_{XX} = 1.5 \times 10\,000/540.6$ = 27.7

$l/r_{YY} = 0.85 \times 10\,000/111.5$ = 76.2

$p_c = 107\ \text{N/mm}^2$

For the allowable stress in bending, see Clause 20, BS449 [1]

$T/t = 35/15 = 2.33; d/t = 1130/15 = 75.3; D/T = 34.2$

$C_s = A = 541.2$ from Table 7, BS449

$p_{bc} = 131\ \text{N/mm}^2$ from Table 8, BS449.

Check the section at the base.

$$f_c = 1561 \times 10^3/427.5 \times 10^2 \qquad\qquad = 36.5 \text{ N/mm}^2$$
$$f_{bc} = 1395.3 \times 10^6 \times 600/12.492 \times 10^9 \qquad\qquad = 67.0 \text{ N/mm}^2$$
$$(f_c/p_c) + (f_{bc}/p_{bc}) = (36.5/107) + (67/131) = 0.34 + 0.51 = 0.84$$

The section is satisfactory.

Check the bearing pressure on the column flange under the crane girder.

Load = 1145 + 50 = 1195 kN

Bearing stress = $1195 \times 10^3/(450 \times 35)$ $\qquad\qquad$ = 75.9 N/mm^2

This is satisfactory.

(b) Building leg 3-4-5
The design actions are:

Eaves — dead + imposed + crane wheels light side + surge:

$P = 120.2 + 114.2 + 2.4 + 2.0$ $\qquad\qquad$ = 234.0 kN

$M = 392.0 + 375.2 - 22.9 + 30.9$ $\qquad\qquad$ = 775.2 kN m.

The trial section and properties are shown in Fig. 6.16(b).

$l/r_{XX} = 1.5 \times 4000/320.2$ $\qquad\qquad$ = 18.7

$l/r_{YY} = 4000/71.5$ $\qquad\qquad$ = 55.9

$p_c = 128.1$ N/mm^2

The allowable stress in bending:

$T/t = 2.5; d/t = 70; D/T = 30$

$C_s = 975$ = A from Table 7 of BS449

$p_{bc} = 155$ N/mm^2 from Table 8 and Table 2 of BS449

$f_c = 234.0 \times 10^3/2.0 \times 10^4$ $\qquad\qquad$ = 11.7 N/mm^2

$f_{bc} = 775.2 \times 10^6 \times 375/2.26 \times 10^9$ $\qquad\qquad$ = 128.6 N/mm^2

$(f_c/p_c) + (f_{bc}/p_{bc}) = (11.7/128.1) + (128.6/155) = 0.09 + 0.83 = 0.92.$

The section is satisfactory.

(c) Roof rafter 5-6-7
Purlins will be spaced at 1.6 m centres. The rafter will be supported on the bottom flange at the third purlin point, i.e. 4.8 m from the eaves.

The design actions are:

Thrust P $\quad = 108.8 + 102.3 + 25.6 + 0.6$ $\qquad = 237.3$ kN

Moment M $\quad = 775.2$ kN m

Maximum shear $= 90.8 + 86.6$ $\qquad = 177.4$ kN.

Try 838 x 292 UB 176. The properties are given in Fig. 6.16(b).

$l/r_{XX} = 1.5 \times 16\ 030/331$ (estimated) $\qquad = 72.6$

$l/r_{YY} = 4800/56.4$ $\qquad = 85.1$

$\quad p_c = 96.9$ N/mm^2

$\quad p_{bc} = 165$ N/mm^2

$\quad f_c = 237.3 \times 10/223.8$ $\qquad = 10.6$ N/mm^2

$\quad f_{bc} = 775.2 \times 10^3/5879$ $\qquad = 131.9$ N/mm^2

$(f_c/p_c) + (f_{bc}/p_{bc}) = (10.6/96.9) + (131.9/165) = 0.11 + 0.80 = 0.91$

Shear stress f_s $\quad = 177.4 \times 10^3/(834.9 \times 14)$ $\qquad = 15.2$ N/mm^2

The section is satisfactory.

(d) Eaves joint 5, 9

The arrangement of the joint is shown in Fig. 6.18(a).

The design actions are:

Moment $M = 775.2$ kN m

Shear V $\quad = 231.6$ kN

Thrust P $\quad = 169.5$ kN

$\Sigma y^2 = 2[200^2 + 700^2 + 1200^2 + 1500^2 + 1650^2] = 13.89 \times 10^6$.

Maximum tension $= \dfrac{[775.2 - 169.5 \times 1.3]\,10^6 \times 1650}{13.89 \times 10^6 \times 10^3}$ $\qquad = 65.9$ kN

Shear assuming uniform distribution:

$S = 231.6/10$ $\qquad = 23.2$ kN.

Try 24 mm HSFG bolts, proof load $\qquad = 207$ kN.

Reduced proof load $= 207 - 1.4 \times 65.9$ $\qquad = 114.7$ kN

Shear value $= 114.7 \times 0.45/1.4$ $\qquad = 36.8$ kN

The joint is satisfactory.

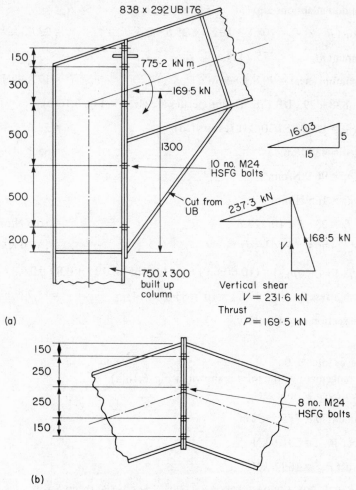

838 x 292 UB 176

150
300
500
500
200

775·2 kN m

169·5 kN

1300

10 no. M24
HSFG bolts

Cut from
UB

750 x 300
built up
column

16·03

5

15

237·3 kN

168·5 kN

V

Vertical shear
$V = 231 \cdot 6$ kN
Thrust
$P = 169 \cdot 5$ kN

(a)

150
250
250
150

8 no. M24
HSFG bolts

(b)

Figure 6.18   (a) Eaves joint. (b) Ridge joint.

(e) Ridge joint

The arrangement of the ridge joint is shown in Fig. 6.18(b). The maximum
moment for design is 291.2 kN m. The thrust is neglected. See Fig. 6.17.

$$\Sigma y^2 = 2[150^2 + 400^2 + 650^2 + 800^2] = 2.49 \times 10^6$$

$$\text{Maximum tension} = \frac{291.2 \times 10^6 \times 800}{2.49 \times 10^6 \times 10^3} \qquad = 93.6 \text{ kN}$$

Use 24 mm HSFG bolts where the allowable load in tension is 124.2 kN

(f) Column base plate

The maximum design conditions are:

Axial load	= 1561 kN
Moment	= 1395.3 kN m.

The allowable stresses are:

Concrete in bearing	= 4.7 kN/mm^2
Bolts in tension	= 120 N/mm^2.

Try a base 1900 mm long x 1200 mm wide.

The arrangement is shown in Fig. 6.19.

$$x = \left[\frac{15 \times 4.7}{(15 \times 4.7) + 120} \right] 1825 \qquad = 675.4 \text{ mm}$$

z = lever arm = $1825 - 675.4/3$ = 1599.9 mm

Take moments about centre line of steel bolts to give:

$M' = 1395.3 + (1561 \times 0.875)$	= 2761.2 kN m
$C = 2761.2/1.599$	= 1726.7 kN
$f_c = 2 \times 1726.7 \times 10^3/(1200 \times 675.4)$	= 4.26 N/mm^2
$T = 1726.7 - 1561$	= 165.7 kN.

Provide five no. 24 mm diameter bolts, net area   = 353 mm^2.

Bolt stress = $165.7 \times 10^3/5 \times 353$ = 93.9 N/mm^2.

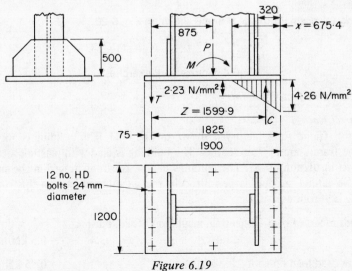

Figure 6.19

The base will be stiffened by plates as shown.

The overhang is 320 mm.

$M = 4.26 \times 320^2/3 + 2.24 \times 320^2/6 = 1.84 \times 10^5$ N/mm

$t = [1.84 \times 10^5 \times 6/185]^{0.5}$ $\quad = 77.1$ mm

Base slab is 80 mm thick.

The stiffener is designed as a cantiliever. A plate 30 mm thick x 1000 mm long x 500 mm wide is adequate. Detail design for the welding is not given. The column flanges and stiffener are connected by full-strength weld to the base. The weld between column flange and stiffener is to be 12 mm fillet weld. These welds will transmit about two-thirds of the flange load to the stiffener.

(g) Deflection
(i) Deflection at crane girder level. Positions 2 and 12
The horizontal deflections are:

Dead load $\quad = 7.88$ mm

Imposed load $= 7.59$ mm.

The full imposed load causes the crane girders to move 15.18 mm apart in a span of 28.8 m. The dead load deflection can be taken up in the fabrication and positioning of the crane girders.
(ii) The wind load deflection at the eaves is 9.88 mm horizontally, i.e. 1/1417 of the height.

The frame is satisfactory with respect to deflection.

(h) The arrangement of the frame is shown in Fig. 6.20

6.4 Coursework exercises

1. *Rigid frame factor building*

(a) *Specification*
The steel frame for a factory is shown in Fig. 6.21. The building is 56 m long and the frames are at 7 m centres. The building is clad with plastic-coated steel sheet on insulation board. The columns in row B can be braced in the end bays only. No lateral support is permitted in the six internal bays in this row. The loading is as follows:

Dead load of roof — sheeting, insulation board, purlins
and rafter $\quad = 0.6$ kN/m^2

Imposed load on roof $\quad = 0.75$ kN/m^2

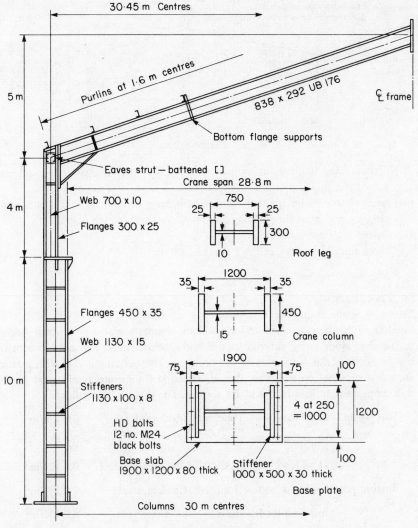

Figure 6.20

Dead load of walls — sheeting, insulation board sheeting
rails and column   = 0.6 kN/m²

Wind load — CP3, Chapter V, Part 2, [2] Location is the North East of England on the outskirts of a city.

(b) *Work required*

(i) Draw framing plans for the building.

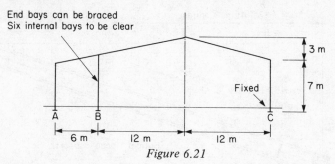

Figure 6.21

(ii) Estimate the loading for an internal portal for the separate load cases and show these on diagrams.
(iii) Analyse the frame for the separate load cases using the I.C.L. Plane-frame program and draw the bending moment diagrams for each case.
(iv) Design the frame and the joints.
(v) Make a detail drawing of the frame and joints.

2. *Rigid-frame warehouse*

(a) *Specification*

The steel frame frame for a two-storey warehouse is shown in Fig. 6.22. The building is 20 m wide × 54 m long and the frames are at 6 m centres. Access to the first floor is by external ramps and internal stairs. The arrangement of the access does not form part of the problem. The building is clad with plastic-coated steel sheet on insulation board. The first floor consists of a cast *in situ* slab supported on floor beams at 2.5 centres. The loading is as follows:

Top portal – Dead load – roof sheeting, insulation, steel   = 0.6 kN/m^2

Imposed load – roof   = 0.75 kN/m^2

Dead load – walls, sheeting, insulation, steel   = 0.6 kN/m^2

Bottom portal – Dead load – floor slab, finishes, steel   = 7.2 kN/m^2

Imposed load – floor   = 8.0 kN/m^2

Dead load – walls: sheeting, insulation steel   = 0.6 kN/m^2

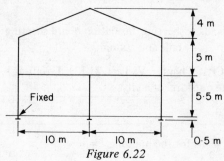

Figure 6.22

Wind load – CP3, Chapter V, Part 2 [2] – location is North East England on the outskirts of a city.

(b) *Work required*

(i) Indicate how you would arrange the frame for manual analysis.
(ii) Draw the steel framing plans for the building.
(iii) Estimate the loading on an internal frame for the separate load cases and show these on diagrams.
(iv) Analyse the frame for the separate load cases using the I.C.L. Plane-frame program and draw the bending moment diagrams for each case.
(v) Design the frame and joints.
(vi) Make a detail drawing of the frame and joints.

References and further reading

[1] BS449, Part 2 (1969). *The use of structural steel in building*. British Standards Institution, London.
[2] CP3, Chapter V, Part 2 (1972). *Loading. Wind Loads*. British Standards Institution, London.
[3] I.C.L. Computer Manual (1969). *Analysis of plane frames and grids* TP.4179 International Computers Limited.
[4] *Steel Designers Manual* (1972). Constrado and Crosby Lockwood, London.
[5] *Structural Steelwork Handbook* (1978). B.C.S.A. and Constrado, London.

7 Rigid design–plastic theory

7.1 Introduction

It is important to include the method of plastic design in the book because this method is now almost exclusively used for the design of single-storey rigid-portal frames. This will also permit a comparison to be made with the elastic method given in Chapter 5.

The plastic method was developed by Lord Baker of Windrush and his research team including Professor M. R. Horne and Professor B. G. Neal at Cambridge University. Practical design methods were presented in various publications by the British Constructional Steelwork Association. These have been replaced by a publication from the Constructional Steel Research and Development Organisation by Morris [6]. Two further B.C.S.A. publications now out of print dealt with stability and the effect of deflections. These are no. 23 by Horne and no. 29 by Horne and Chin. The charts from there have been included in the Constrado publication *Plastic design* – Supplement [2].

The method of plastic design outlined below has been based on these Constrado publications. This is presented here with their kind permission. The treatment is restricted to single-bay single-storey portal frames subjected to building dead loads, imposed loads on the roof and wind loads. Practical design examples are also given.

7.2 Plastic analysis

7.2.1 Basic principles

The plastic hinge is the central concept in plastic analysis. The behaviour of a section in bending is based on the idealized stress–strain curve for structural steels shown in Fig. 7.1(a). The formation of the hinge is shown in Fig. 7.1(b) on the figure. Here outer fibres on reaching the yield stress continue to strain at constant stress until yielding reaches the neutral axis. The beam section is then fully plastic and acts as a hinge under the plastic moment, M_p.

Here the plastic analysis of rectangular and pitched roof pinned and fixed base portals are considered. As the load on the frame is increased hinges form

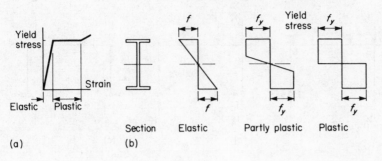

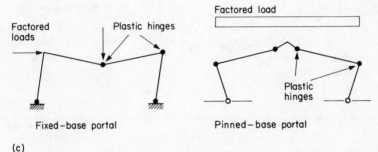

Figure 7.1 (a) Stress-strain curve. (b) Formation of the plastic hinge. (c) Collapse mechanisms.

successively starting at points of maximum elastic moment. Collapse occurs when sufficient plastic hinges have formed to convert the frame into a mechanism. In general, the number of hinges required is one more than the statical indeterminacy. However, with vertical loading, collapse occurs in the beam or rafter and more hinges than this form. See Fig. 7.1(c).

For the location of the hinges to be correct, the plastic moment at the hinges must not be exceeded at any other point in the frame. The critical mechanism is the one which gives the lowest value for the collapse or factored load, where

Collapse or factored load = working load x load factor λ

BS449 states in Clause 9a that an adequate value of the load factor is to be used. Values used here are taken from the Constrado publication by Morris and Randell [6]. These are:

Dead and imposed load   λ = 1.7

Dead, imposed and wind load   λ = 1.36.

7.2.2 Plastic analyses for pinned base portals

(a) Rectangular pinned base portal
The rectangular portal and typical factored loading is shown in Fig. 7.2(a). The

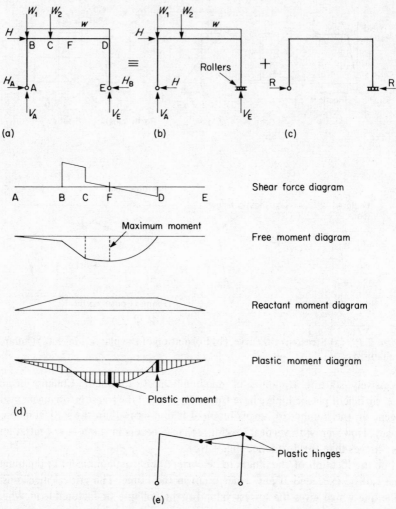

Figure 7.2 (a) Frame and loads. (b) Released structure. (c) Redundant action. (d) Steps in plastic analysis. (e) Collapse mechanism.

collapse mechanism which occurs depends on the location and relative magnitude of the various loads. One possible case is considered. The frame has a statical redundancy of one. So in general two hinges must form to cause collapse.

The plastic analysis is carried out in the following stages:

(i) the frame is released to a statically determinate state by inserting rollers at one support, Fig. 7.2(b);

(ii) the free bending moment diagram is drawn. In this case the position of the maximum sagging moment in the beam is found by drawing the shear force diagram;

(iii) the reactant bending moment diagram due to the redundant horizontal reaction is drawn;
(iv) the free and reactant bending moment diagrams are combined to give the plastic bending moment diagram with sufficient hinges to cause the frame to collapse. The plastic bending moment must not be exceeded at any point. The plastic hinges are located by inspection.

The various moment diagrams are shown in Fig. 7.2(d). Two hinges form at F and D and the collapse mechanism is shown in Fig. 7.2(e).

If vertical loads only are applied, collapse occurs in the beam with three hinges forming. The loading, collapse diagram and plastic bending moment diagram for this case are shown in Fig. 7.3(a). When the horizontal load is of much greater

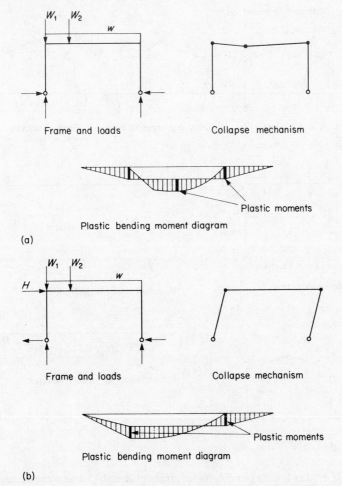

Figure 7.3 (a) Vertical loads only on the frame. (b) Large horizontal load.

196   Steel structures

significance than the vertical loads, sideways collapse occurs as shown in Fig. 7.3(b).

(b) Pitched roof pinned base portals
The analysis may be carried out in the same manner as that outlined for the rectangular portal set out above. The portal and general factored loading are shown in Fig. 7.4(a). This also shows the collapse mechanisms for the following cases:

(i)   dead and imposed load;
(ii) dead, imposed and wind load.

The plastic analysis for the second load case is shown in Fig. 7.4(b).

The exact location of the hinge near the apex may be found either by successive

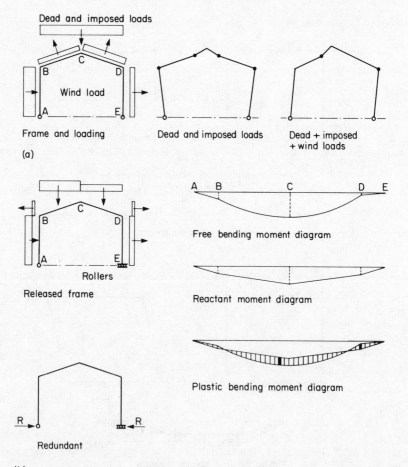

Figure 7.4   (a) Loading and collapse mechanisms. (b) Steps in the plastic analysis for dead + imposed + wind load.

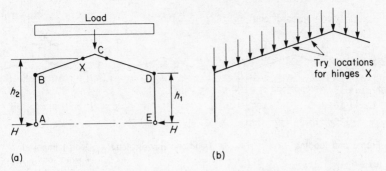

Figure 7.5 (a) Collapse mechanism. (b) Loads applied at purlin points.

trials or mathematically if the roof loading is taken as uniformly distributed. See Fig. 7.5(a). Alternatively, the roof loading may be taken as applied at the purlin points, as shown in Fig. 7.5(b) when the hinge will occur at a purlin point and these may be checked in turn to see which gives the maximum value for the plastic moment. Referring to Fig. 7.5(a), equating expressions for the plastic moment at the hinges gives

$$Hh_1 = M_x - Hh_2 = M_p$$

where M_x is the free moment at X. Solve for H and determine the plastic moment M_p.

7.2.3 Plastic analysis for fixed base portals

The analysis for a pitched roof portal only is discussed. The analysis for a rectangular portal or single-bay portal of other shape follows lines similar to those set out here. A fixed base portal and general loading is shown in Fig. 7.6(a). The collapse mechanisms for symmetrical dead and imposed loading and asymmetrical dead, imposed and wind loading are shown in the figure. Note that in some cases for dead, imposed and wind load the hinge may form at A instead of B. For the fixed base portal it is convenient to make the frame statically determinate by dividing it into two cantilevers at the apex C as shown in Fig. 7.6(b). The free moments may be calculated and the free moment diagram drawn. There are three redundants at the release, the moment M, shear S, and thrust R. The reactant moment diagram can be drawn. The free and reactant diagrams are combined to give the plastic bending moment diagram.

The values of the three redundants M, R and S and the plastic moment may be found by forming the equations for the plastic moments at the hinges in terms of the free and reactant moments and solving these. The location of the hinge at X must be found by trial and it is convenient to consider the loads applied at the purlin points. Then purlin points near the apex may be checked in turn to see which gives the maximum value for the plastic moment.

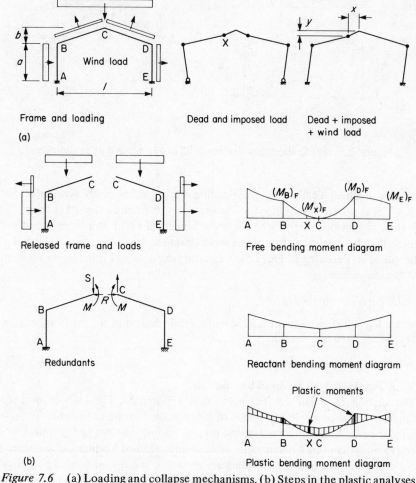

Figure 7.6   (a) Loading and collapse mechanisms. (b) Steps in the plastic analyses.

The equations for the plastic moments at the hinges are:

B $M_p = (M_B)_F - [M + Rb - Sl/2]$

X $M_p = M + Ry - Sx - (M_X)_F$

D $M_p = (M_D)_F - [M + Rb + Sl/2]$

E $M_p = M + R(a + b) + Sl/2 - (M_E)_F,$

where $(M_B)_F$, $(M_X)_F$, $(M_D)_F$, $(M_E)_F$ are the free moments at B, X, D and E, respectively. Solve for M, R, S and M_p. The process is shown in the example that follows.

7.2.4 Haunches

Haunches may be provided at the eaves and apex of the frame. The haunch at the apex has no effect on the strength as the hinge always forms some distance away from the apex. The haunch at the eaves causes the hinge to form in the column at the bottom of the haunch. This reduces the value of the plastic moment and so the section required will be lighter than that required for a frame with no haunch.

The collapse mechanism for a frame with haunches is shown in Fig. 7.7(a) for the case of symmetrical dead and imposed load. For a given depth of haunch, a, the equations for the plastic moments at A, Y and X can be formed and solved to give the redundants M and R at C and the plastic moment M_p. The haunch is shown in Fig. 7.7(b). The length of the haunch in the rafter must be such that the haunched portion remains in the elastic state. Thus the moment at the end of the haunch must not exceed

$$M_w = f_y Z = f_y S / 1.15 = 0.87 M_p,$$

where Z is the elastic modulus, S the plastic modulus, $S/Z = 1.15$ – shape factor for rolled I sections, and f_y the yield stress.

Provision of a haunch is essential if a bolted joint with high-strength friction grip bolts is to be used. This is required to give sufficient lever arm for the bolts to be designed. Joint design is given in the design example in Section 7.4.5.

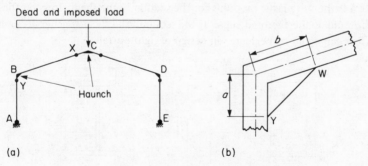

Figure 7.7   (a) Collapse mechanism. (b) Haunch.

7.3 Plastic design

7.3.1 Beams

A symmetrical beam section subjected to a plastic moment M_p is chosen so that its plastic moment of resistance is equal to or exceeds M_p. This is achieved by selecting a beam with plastic modulus S slightly greater than that given by

$$S = M_p / f_y.$$

Values of the plastic modulus of section for rolled sections are given in the B.C.S.A. *Structural Steelwork Handbook* [8].

Referring to Fig. 7.8(a), the plastic modulus for a rectangular section $b \times d$ is $S = bd^2/4$. By definition, the plastic modulus is the algebraic sum of the first moments of area about the equal area axis. Thus for the symmetrical I section shown in Fig. 7.8(b)

$$S = BT(D - T) + td^2/4$$

where B is the width of flange, D the overall depth of beam, d the web depth between flanges, T the flange thickness, and t the web thickness.

The yield stress f_y depends on the grade of steel and thickness of material and is given in BS4360 [11]. For example, for Grade 43 steel for rolled I sections

Thickness up to 16 mm $f_y = 255 \text{ N/mm}^2$

Thickness over 16 mm and up to 40 mm $f_y = 245 \text{ N/mm}^2$.

If the beam section is subjected to shear force as well as the plastic moment, the shear force reduces the plastic moment of resistance of the section. It is assumed that the web only resists shear so that portion of the plastic modulus contributed by the web is reduced. The reduced plastic modulus is given by:

$$\text{Reduced } S' = S - \frac{td^2}{4} \{1 - [1 - 3(q/f_y)^2]^{1/2}\}$$

where S is the full plastic modulus for the section, q the shear stress on the web $= V/td$, and V the factored shear at the section. The shear is generally small and its effect can often by neglected in design of light portals.

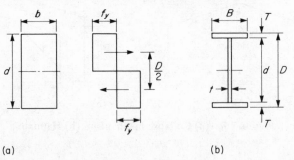

(a)   (b)

Figure 7.8   (a) Rectangular section. (b) Symmetrical I section.

7.3.2 Column design

The column section is subjected to a factored axial load P in addition to the plastic moment M_p. The axial load reduces the plastic moment of resistance of the section. The plastic analysis is as follows.

(i) The plastic stress distribution is made up of two parts, one of which resists moment and one axial load.

(ii) Bending is resisted by two equal areas extending inwards from the extreme fibres. The central area resists direct load. This area may be in the web only, or extend into the flanges. It reduces the moment of resistance of the section. The stress distribution is shown in Fig. 7.9.

Formulae for calculating the reduced values of the plastic moduli of section are given in the B.C.S.A. *Structural Steelwork Handbook* [8]. These depend on values of

n = the mean axial stress f_c/yield stress f_y,

where f is the factored load P/area of section A. Change values of n are given for each section. For

(i) lower values of n, the neutral axis lies in the web,
(ii) upper values of n, the neutral axis lies in the flange.

A separate formula is given for each case.

If the column section is subjected to a significant shear force as well as the plastic moment and axial load, this will also reduce the moment of resistance. This will have to be taken into account as set out in Section 7.3.1 for beam design.

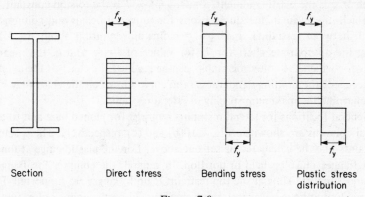

| Section | Direct stress | Bending stress | Plastic stress distribution |

Figure 7.9

7.3.3 Section stability
The section must be such that a plastic hinge can form without buckling occurring in the flanges or web. Considerations here are: (a) flanges, and (b) web.

(a) Flanges
The possibility of buckling occurring in the flanges is eliminated by limiting the flange width to thickness ratio. Referring to Fig. 7.8(b), the limiting ratio B/T is 13.75 for Grade 43 steel. Sections that are unsatisfactory with respect to flange buckling are noted in a table in the B.C.S.A. *Structural Steelwork Handbook* [8].

(b) Web
With regard to web buckling, if the web depth to thickness ratio d/t exceeds

limiting values depending on the grade of steel, e.g. 53 for Grade 43 steel, premature web buckling occurs if the mean axial stress exceeds a certain value related to the section size. No universal column sections are unsuitable. Some universal beam sections when used as columns are unsatisfactory. These are listed in the table in the B.C.S.A. *Structural Steelwork Handbook* [8].

7.3.4 Column stability
The frame must be restrained laterally to prevent buckling of the member or compression flange. Restraints are needed near plastic hinges and in the elastic regions between hinges on long columns.

Charts to check column stability are given in the Constrado publication, *Plastic Design* [2]. This incorporates the B.C.S.A. publication no. 23. These charts were prepared by Professor M. R. Horne. The method given is outlined briefly.

The member stability depends on the torsional stiffness T, the slenderness ratio l/r_{YY} and the ratio and direction of the end moments acting on the column. Consider the column shown in Fig. 7.10(a) which is subjected to the factored load P and moments M'_X and M''_X acting about the major axis of which M'_X is the plastic moment. $\beta = M''_X/M'_X$, T is the torsion constant, values of which are given in the publications for universal beams and columns, l the length between restraints, and r_{YY} = radius and gyration about the YY axis. Using the appropriate chart for T, for values of l/r_{YY} and β, the mean axial stress for which the column is stable, can be read off the chart. If the axial stress is exceeded, the limiting slenderness ratio curve shown on the chart may be used to determine the maximum spacing of lateral restraints.

General locations for lateral restraints required for pinned base and fixed base portal columns are shown in Fig. 7.10(b) and (c), respectively. The spacing S of restraints may be calculated as set out above. For the plastic hinge at the eaves, both flanges must be held in position. In general, the compression flange only requires support. Thus at the top restraint R both flanges are supported. For the fixed column at the lower restraint a sheeting rail on the compression flange will provide sufficient restraint. The reader should also consult Horne and Ajmani [9], and Horne [10].

Lateral restraints can be in the form of latticed members or stays from sheeting rails as shown for the elastic design in Fig. 5.19. The Constrado publication recommends that the cross-sectional area of the restraint member should be at least 4% of the area of the member supported.

It is also necessary to check that the length between restraints RA or RQ, Fig. 7.10(b) or (c) respectively, remains elastic and stable under the factored loads. The method given in the B.C.S.A. publication no. 23 and the charts in the *Plastic Design – Supplement* [2] can be used to do this. See Fig. 7.10(a) where in this case a column of length l carries factored end moments M'_X and M''_X about the major axis and a factored axial load P. The moment M'_X is greater than M''_X. For values of $\beta = M''_X/M'_X$ a moment coefficient m is read from chart 33/23.

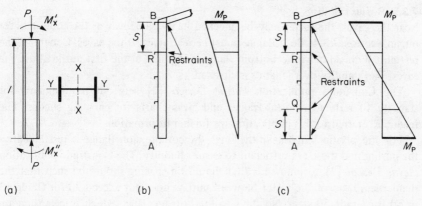

Figure 7.10 (a) Column loads and moments. (b) Pinned-base column. (c) Fixed-base column.

Then the equivalent uniform major axis moment is $M_X = mM'_X$. If A is column area, Z_X the elastic modulus for the major axis and r_{XX} and r_{YY} the radii of gyration about the major and minor axis, respectively, then

$$p = P/A \qquad = \text{axial stress}$$

$$f'_X = M'_X/Z_X = \text{bending stress at the top}$$

$$f_X = M_X/Z_X = \text{bending stress due to the equivalent moment}$$

$$l/r_{XX} \qquad\qquad = \text{slenderness ratio for the major axis}$$

$$l/r_{YY} \qquad\qquad = \text{slenderness ratio for the minor axis}$$

$$T \qquad\qquad = \text{torsional constant for the section}$$

$$p' \qquad\qquad = p + f_X^2/T.$$

From Chart 34/23 using l/r_{XX} and p read off N_X.

From Chart 35/23 for Grade 43 steel using l/r_{YY} and p' read off f. Then if f_Y is the yield stress for the material:

(a) the yield stress at the top is not exceeded if

$$p + f'_X \quad \leqslant f_Y;$$

(b) the yield stress near the centre of the column is not exceeded if

$$p + N_X f_X \leqslant f.$$

Note that the column lengths between restraints RA or RQ (Fig. 7.10b or c) may also be checked using elastic theory from BS449 for the unfactored loads and moments.

7.3.5 Rafter stability

Near the eaves, the rafter may be checked in the same way as the column as set out in Section 7.3.4. Both flanges must be restrained at the eaves. Generally one restraint with stays to the bottom flange is required at the first purlin above the eaves. The stability over length s is checked.

The Constrado publication *Plastic Design* [2] sets out requirements for stability of rafters in portal frames and gives charts for carrying checks. The reader is referred to this publication for further information.

For the plastic hinge near the ridge the compression flange is restrained by the purlins and these are sufficient to ensure stability. The Constrado publication *Plastic Design* [2] recommends that the purlin spacing should be such that the slenderness ratio of the rafter between purlins does not exceed 70 for Grade 43 or 60 for Grade 50 steel. No stays to the bottom flange which is in tension are required. The reader should also consult Horne and Ajmani [9] and Horne [10].

If the portal rafter is haunched the plastic hinge forms in the column and the haunch is made sufficiently long so that the rafter remains elastic in the portion near the eaves. Elastic criteria can be used to locate lateral supports. Again, restraints to the bottom flange are required at first purlin above the eaves.

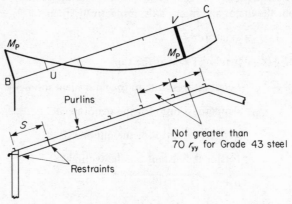

Figure 7.11

7.3.6 Correction for deflection

In the plastic analysis set out above the deflections are assumed to be small and are not considered in the analysis. In slender pitched roof frames, the change in frame geometry due to the outward deflection at the eaves reduces the strength of the frame and so the load factor.

Charts for determining the reduction are given in the Constrado publication *Plastic Design* [2] which incorporates the B.C.S.A. publication no. 29. Charts are given for pinned base and fixed base portals for various values of roof slope. Using values of effective slenderness ratio CL/d and the span-to-height to eaves ratio L/h_1, values of the reduction in load factor can be read from the chart.

Here:

C has values of 1/48 for grade 50 steel and 1/61 for grade 43 steel, L is the span (mm), d the depth of the frame section (mm) and h_1 the height to the eaves (mm). The actual load factor for the section used is reduced and this should not be less than the minimum load factor required.

The charts were produced for grade 50 steel. They can be used for grade 43 steel but should not be used for grade 55 steel. Pinned base portals, in particular, require checking. The reduction in load factor for these portals is limited to 10%. Plastic design should not be used if this limit is exceeded.

7.3.7 Deflections at working loads
Charts are given in the publications referred to in Section 7.3.6 to determine the horizontal deflection at the eaves due to working loads. The charts cover pinned and fixed base portals with various slopes for the rafters.

Plastic design – single-storey building

7.4.1 Specification
Redesign the pinned based portal for the building specified in Section 5.6 using plastic theory. The frame will be analysed for the two cases of dead and imposed load and dead, imposed and wind load. The frame dimensions and working loads for the various load cases are shown in Fig. 7.12. The wind loads on the roof have been resolved vertically and horizontally. The load factors to be applied are

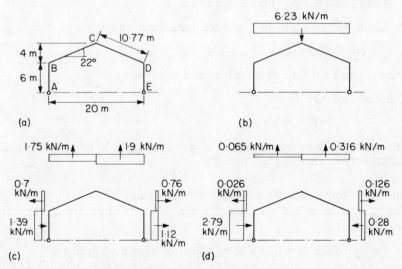

Figure 7.12 (a) Frame and dimensions. (b) Dead and imposed loads. (c) Wind – internal pressure. (d) Wind – internal suction.

as follows:

Dead and imposed loads $\lambda = 1.7$

Dead, imposed and wind loads $\lambda = 1.36$.

Use Grade 43 steel with a yield stress of 255 N/mm^2 for sections up to 16 mm thick.

7.4.2 Analyses

(a) Dead and imposed load
Release the frame by inserting rollers at E as shown in Fig. 7.13(a). Calculate the free moment at C and draw the free moment diagram

Free $M_C = 6.23 \times 20^2/8$ $\qquad\qquad\qquad\qquad\qquad$ = 311.5 kN m.

The free moment diagram is shown in Fig. 7.13(b). The reactant moment diagram due to the redundant R is shown in Fig. 7.13(c). The collapse mechanism is shown in Fig. 7.13(d) and the value of x must be determined to give a maximum value of the plastic moment m_p due to the unfactored loads and ensure that this is not exceeded at any point in the frame.

Free $M_X = 62.3x - 3.115x^2$.

Equating expressions for the plastic moments at B and X gives the equation

$6H = 62.3x - 3.115x^2 - H(6 + 0.4x)$.

Re-arrange the equation to give

$H = (62.3x - 3.115x^2)/(12 + 0.4x)$.

Put $dH/dx = 0$, collect terms to give the quadratic equation

$x^2 + 60x - 600 = 0$.

Solve to give $x = 8.73$ m or 9.4 m measured on slope.

This gives $H = 19.78$ kN

$\qquad\qquad m_p = 118.71$ kN.

The plastic bending moment diagram is shown in Fig. 7.13(e). Factored plastic moment $M_p = 118.71 \times 1.7 = 201.81$ kN m.

(b) Dead + imposed load + wind load internal pressure
The released frame and loading are shown in Fig. 7.14(a). The reactions and moments are:

$R_E = [(15.06 \times 3) + (0.24 \times 8) + (43.3 \times 15) + (44.8 \times 5)]/20$

$\qquad\qquad\qquad\qquad\qquad\qquad\qquad\qquad$ = 46.03 kN

$M_B = -(15.3 \times 6) + (8.34 \times 3)$ $\qquad\qquad\qquad\qquad$ = $-$ 66.78 kN m

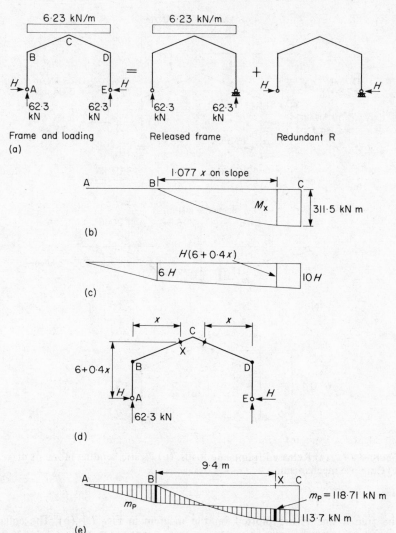

Figure 7.13 (a) Frame and release. (b) Free bending moment diagram. (c) Reactant moment diagram. (d) Collapse mechanism. (e) Plastic bending moment diagram.

$M_D = -6.72 \times 3$ $= -20.16$ kN m

$M_C = -(46.03 \times 10) - (6.72 \times 7) - (3.04 \times 2) + (43.3 \times 5) = -296.92$ kN m.

The free bending moment at x from B is

$$M_x = -15.3(6 + 0.4x) - 42.07x + 8.34(3 + 0.4x) - 0.7(0.4x)^2/2 + 4.48x^2/2$$

$$= -66.78 - 44.854x + 2.184x^2.$$

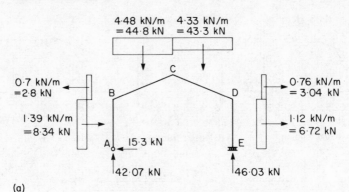

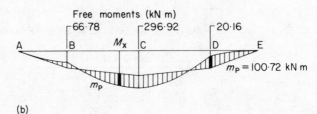

(a)

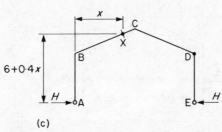

(b)

(c)

Figure 7.14   (a) Released frame and loads. (b) Plastic bending moment diagram. (c) Collapse mechanism.

The free moments are plotted on the diagram in Fig. 7.14(b). The collapse mechanism is shown in Fig. 7.14(c) with hinges at X and D. Equating expressions for the plastic moments at D and X gives the equation

$$6H - 20.16 = 66.78 - 44.854x - 2.184x^2 - H(6 + 0.4x).$$

Rearrange the equation to give

$$H = (86.94 + 44.834x - 2.184x^2)/(12 + 0.4x).$$

Put $dH/dx = 0$ and collect terms to give the quadratic equation

$$x^2 + 60.04x - 576.4 = 0.$$

Solve to give $x = 8.42$ m.

This gives $\qquad$ $H = 20.15$ kN,

$$m_p = (20.15 \times 6) - 20.16 \qquad\qquad = 100.72 \text{ kN m.}$$

The plastic bending moment diagram is shown in Fig. 7.14(b). The factored moment $M_p = 100.72 \times 1.36 \qquad\qquad = 136.97$ kN m.

(c) Dead + imposed + wind load internal suction

The analysis similar to that in (b) above gives:

$H = $ 20.16 kN,

$m_p = 126.0$ kN m.

The factored moment is 171.3 kN m.

The design is made for dead and imposed load where the factored moment is 201.81 kN m.

7.4.3 Frame design

(a) Column design

Factored moment M_p = 201.81 kN m.

Plastic modulus required to resist moment only

$S = 201.8 \times 10^3/255 = 791.4$ cm^3.

Try 406 x 140 UB 46, the properties of which are

$S = 888.4$ cm^3, $Z = 777.8$ cm^3, $r_{YY} = 3.02$ cm,

$A =$ 59.0 cm^3, $r_{XX} = $ 16.29 cm.

Actual load factor $= 888.4 \times 255/118.71 \times 10^3 \qquad\qquad = 1.91$.

Check the reduction in plastic moment of resistance caused by the axial load for a load factor of 1.91

Axial stress $p = 1.91 \times 62.3 \times 10/59$	$=$ 20.2 N/mm^2.
$n = p/f_y = 20.2/255$	$=$ 0.079.
Reduced plastic modulus $= 888.4 - 1258n^2$	$= 880.5$ cm^3.
Reduced load factor $= 880.4 \times 1.91/888.4$	$=$ 1.89.

The reduction due to axial load is negligible. Note that stability checks are made and joints designed for a load factor of 1.91 so that all parts have the same strength.

Check the column for stability. Provide four sheeting rails spaced as shown in Fig. 7.15(a). Consider the top portion BF, where the moment at F is 96.95 kN/m.

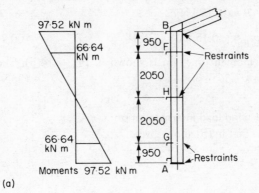

(a)

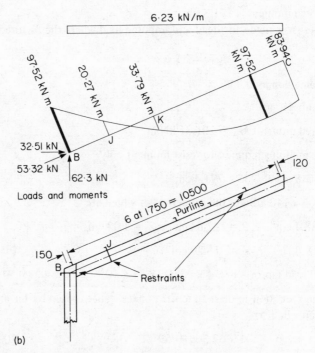

(b)

Figure 7.15   (a) Column. (b) Rafter.

Torsion constant $T = 152 \text{ N/mm}^2$ Table C – Plastic Design [2]

Axial stress $p = 20.2 \text{ N/mm}^2$

Moment ratio $\beta = 96.95/118.71 = 0.82$

From Chart 2/23 [2] the limiting slenderness ratio l/r_{YY} = 39.

Allowable spacing of restraints = 39 × 30.2   = 1178 mm

The spacing selected is satisfactory. Provide restraints to both flanges at the eaves B and at F.

Check the stress at F under the factored loads.

Axial load = 62.3 + 0.48 x 5 x 1.1	= 64.9 kN
Factored axial stress p = 1.91 x 69.9 x 10/59	= 21.02 N/mm^2
Factored bending stress f'_X = 1.91 x 96.95 x 10^3/777.8	= 238.1 N/mm^3
Total stress $p + f'_X$ = 259.12 N/mm^2	> 255 N/mm^2

This exceeds the yield stress f_y. Treat portion FG as plastic in the same way as BF above.

Provide restraints to both flanges at G.

Check Portion GA to show that this remains elastic and stable under the factored loads.

Axial load = 62.3 + 0.48 x 5 x 2.2	= 67.6 kN
Moment	= 75.18 kN
Factored axial stress p = 1.91 x 67.6 x 10/59	= 21.87 N/mm^2
Factored bending stress f'_X = 1.91 x 75.18 x 10^3/777.8	= 184.6 N/mm^2
Moment ratio β	= 0

From Chart 33/23 [2] m = 0.565.

Bending stress due to the equivalent moment

f_X = 0.565 x 184.6	= 104.3 N/mm^2
p' = 21.87 + 104.3^2/152	= 93.4 N/mm^2.
Slenderness ratios l/r_{XX} = 1.5 x 6000/162.9	= 55.2
l/r_{YY} = 3800/30.2	= 125.8.

From Chart 34/23 [2] for l/r_{XX} = 55.2, p = 21.87 N/mm^2; read off

$$N_X = 1.04.$$

From Chart 35/23 [2] for l/r_{YY} = 125.8, p' = 93.4 N/mm^2, read off

$$f = 183 \text{ N/mm}^2.$$

Then the checks are:

Top: $p + f'_X$ = 21.87 + 184.6 = 206.5 $< f_y$ = 255 N/mm^2

Centre: $p' + N_X f_X$ = 21.87 + 1.04 x 104.3 = 130.3 $< f$ = 183 N/mm^2.

The column is satisfactory as shown with restraints at the eaves and the top two sheeting rails F and G.

(b) Rafter design

Provide seven purlins spaced at 1750 mm as shown in Fig. 7.15(b). The moments at the purlin points are shown in the figure. Check portion BH.

Moment ratio	$\beta = 32.47/118.71$	$= 0.274$
Axial stress	$p = 1.91 \times 41.5 \times 10/59$	$= 13.4 \text{ N/mm}^2$

From Chart 2/23 [2] the slenderness ratio l/r_{YY} $= 80$.

Allowable spacing of restraints $= 80 \times 30.2$ $= 2416$ mm

Spacing provided $= 1900$ mm

Provide restraints to the bottom flange at the purlin at H. The purlins provide restraints to the compression flange for the hinges near the apex. No stays to the bottom flange are required

(c) Correction for deflection

$CL/d = 20/(61 \times 0.402)$ $= 0.816$

$L/h_1 = 20/6$ $= 3.33$

Roof slope $\phi = 22°$

From Chart 10/29 [2] the correction = 10%

The corrected load factor $= 1.89 - 10\% = 1.7$

This is satisfactory.

(d) Working load deflection

From Chart 14/29 [2] for $\phi = 22°$ and $L/h_1 = 3.33$

$\delta/[h_1 L\sigma_y/d\lambda_b] 10^6$ $= 0.83$

where $h_1 = 6\,000$ mm

$L = 20\,000$ mm

$d = 402.3$ mm

σ_y = yield stress $= 255 \text{ N/mm}^2$

λ_b = load factor according to simple plastic theory $= 1.91$.

Then $\delta = \dfrac{0.83 \times 6\,000 \times 20\,000 \times 244}{402.3 \times 1.91 \times 10^6}$ $= 33.1$ mm

$\delta/h_1 = 33.1/6000$ $= 1/181$.

The cladding will reduce this deflection.

(e) Conclusion

The section selected 406 x 140 UB 46 is satisfactory. Restraints are required to both flanges at the eaves, on the column at the top two sheeting rails below the eaves and on the rafter at the second purlin above the eaves. See Fig. 7.15.

7.4.4 Redesign of frame with haunches at eaves

A haunch of depth 500 mm is shown in Fig. 7.16(a). This will permit a bolted joint with high-strength friction grip bolts to be used at the eaves. The collapse mechanism with the hinge in the column at the end of the haunch is shown in Fig. 7.16(b). The design is made for the case of dead and imposed load. Equating the expressions for plastic moments at X and Y gives

$$5.5H = 62.3x - 3.115x^2 - H(6 + 0.4x)$$

$$H = (62.3x - 3.115x^2)/(11.5 - 0.4x).$$

Put $dH/dx = 0$ and collect terms and reduce to give:

$$x^2 + 57.5x - 575 = 0.$$

Solve to give $x = 8.69$ m.

This gives $H = 20.45$ kN,

$$m_p = 112.5 \text{ kN m}.$$

The plastic bending moment diagram is shown in Fig. 7.16(c).

Factored moment = 1.7 x 112.5   = 191.2 kN m.

Theoretically a lighter section than that needed for the frame of uniform section can be used. However, in this case, the section could not be made lighter. The load factor on the next lighter section is not adequate when this is corrected for the effect of deflection.

The length of the haunch in the rafter must be sufficient to ensure that the stress in the rafter at the end of the haunch does not exceed the yield stress. The moment at the end of the haunch, see Fig. 7.15(b)

$$M = 118.71 + 6.23 \times 0.65^2/2 + 19.78 \times 0.65 \times 4/10 - 62.3 \times 0.65$$

= 84.67 kN m.

This is not to exceed $0.87m_p$ = 103.2 kN m.

The haunch length is satisfactory.

7.4.5 Design of joints

(a) Eaves joint

The arrangement for the eaves joint is shown in Fig. 7.16(a). Assuming 20 mm diameter HSFG bolts the plastic moment of resistance of the top four bolts in

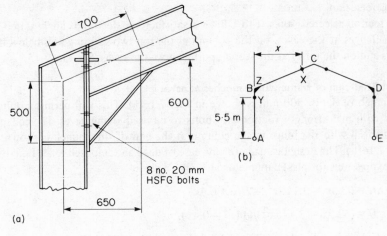

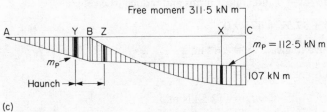

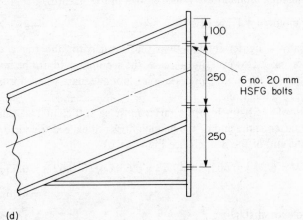

Figure 7.16 (a) Eaves joint. (b) Collapse mechanism. (c) Plastic bending moment diagram. (d) Ridge joint.

the joint is given by

$$M_R = 4 \times 0.6 \times 144 \qquad\qquad = 345.6 \text{ kN m},$$

where the proof load of one bolt is 144 kN and the lever arm of the group is

0.6 m. The factored moment at the eaves = 1.91 x 118.71 = 226.74 kN m. The axial load in the rafter has been neglected.

The joint is satisfactory.

(b) Ridge joint
The arrangement for the ridge joint is shown in Fig. 7.16(d). The moment of resistance of joint with lower two bolts at the proof load is

$$M_R = 2[0.6 + 0.35^2/0.6 + 0.1^2/0.6]\,144 \qquad\qquad = 236.3 \text{ kN m.}$$

The factored moment at ridge = 1.91 x 113.6 $\qquad\qquad$ = 216.9 kN m.

The joint is satisfactory.

7.5 Redesign of the portal using fixed bases

7.5.1 Plastic analysis
The design is made for dead and imposed load as this combination gives the maximum design conditions. The frame is statically determinate by making a cut at the ridge C. Owing to symmetry only one half of the frame need be considered. The released half frame and loading are shown in Fig. 7.17(a). There are two redundant actions at the cut, the moment M and thrust H. The free moments are calculated and the free moment diagram is drawn in Fig. 7.17(c).

The collapse mechanism is shown in Fig. 7.17(b) with hinges at the base, eaves and near the ridge. The location of the hinge near the ridge must be found by successive trials. The calculations are given for a hinge at 2.1 m from the ridge. This location gives the maximum value for the plastic moment. The equilibrium equations for the hinges are

A $\quad m_p = M + 10H - 311.5$

B $\quad m_p = -M - 4H + 311.5$

P $\quad m_p = M + 0.84H - 13.73.$

Solving these equations gives the values

$$H = 32.51 \text{ kN}$$

$$M = 83.92 \text{ kN m}$$

$$m_p = 97.52 \text{ kN m.}$$

The plastic bending moment diagram is shown in Fig. 7.17(c).

The factored moment $m_p = 97.32 \times 1.7 \qquad\qquad$ = 165.8 kN m.

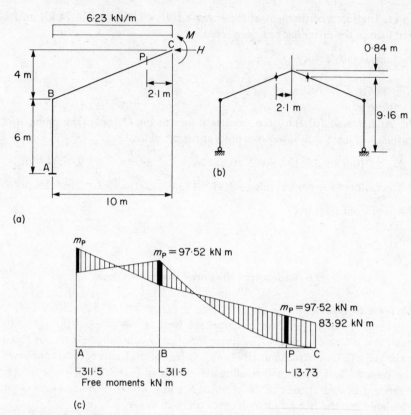

Figure 7.17   (a) Release frame and loads. (b) Collapse mechanism. (c) Plastic bending moment diagram.

7.5.2 Frame design

(a) Column design

Plastic modulus required to resist the moment only

$$S = 165.8 \times 10^3/255 \qquad\qquad = 650.2 \text{ cm}^3.$$

Try 406 x 140 UB 39.

$$S = 720.8 \text{ cm}^3, \quad Z = 626.9 \text{ cm}^3, r_{YY} = 2.89 \text{ cm},$$

$$A = 48.4 \text{ cm}^2, \quad r_{XX} = 15.88 \text{ cm}.$$

Actual load factor = $720.8 \times 255/97.52 \times 10^3$ = 1.88.

Check the reduction in plastic moment of resistance caused by the axial load at the base

Axial load = $62.3 + (0.48 \times 5 \times 6)$ = 76.7 kN.

Factored axial stress $p = 1.88 \times 76.7 \times 10/48.4$ $= 29.8 \text{ N/mm}^2$

                 $n = 29.8/255$ $= 0.116$

Reduced plastic modulus $= 720.8 - 964.9n^2$ $= 707.8 \text{ cm}^2$.

Reduced load factor $= 707.8 \times 1.88/720.8$ $= 1.85$.

The reduction due to axial load is negligible.

Check the column for stability. Provide four sheeting rails as shown in Fig. 7.18(a). Consider the lower portion GA where the moment at G is 66.64 kN m and lateral restraints are provided at F and G as well as at the eaves.

Torsion constant $T = 110 \text{ N/mm}^2$

Axial stress $p = $ 29.8 N/mm^2

Moment ratio $\beta = $ $66.64/97.52 = 0.683$

From Chart 1/23 [2], the limiting slenderness ratio curve gives

      $l/r_{YY} = 34$

Allowable spacing of restraints $= 34 \times 28.9 = 982.5$.

The spacing selected is satisfactory. This is 950 mm.

Check the portion of the column FG between restraints to show that this remains elastic and stable.

Load $G = 62.3 + 0.48 \times 5 \times 5.05$ $= 74.42 \text{ kN}$

Moment $= 66.64 \text{ kN m}$

Factored axial stress $p = 1.88 \times 74.42 \times 10/48.4$ $= 28.9 \text{ N/mm}^2$

Factored bending stress $f'_X = 1.88 \times 66.64 \times 10^3/626.9$ $= 199.8 \text{ N/mm}^2$

Moment ratio β $= -1$

From Chart 33/23 [2] $m = 0.391$.

Bending stress due to the equivalent moment:

$f_X = 0.391 \times 199.8$ $= 78.12 \text{ N/mm}^2$

$p' = 28.9 + 78.12^2/110$ $= 84.38 \text{ N/mm}^2$.

Slenderness ratios: $l/r_{XX} = 1.5 \times 6000/158.8$ $= 56.67$

                $l/r_{YY} = 4100/28.9$ $= 141.86$.

From Chart 34/23 [2] for $l/r_{XX} = 56.67$, $p = 28.94 \text{ N/mm}^2$; read off

                             $N_X = 1.06$.

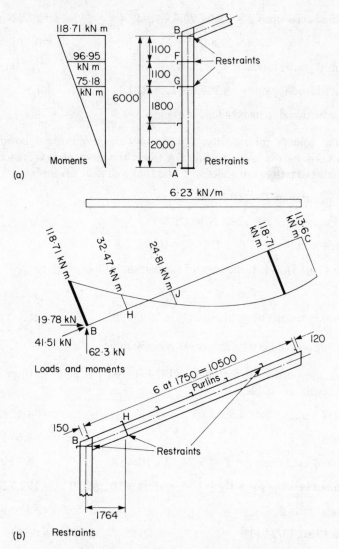

Figure 7.18   (a) Column. (b) Rafter.

From Chart 35/23 [2] for $l/r_{YY} = 141.86$; $p' = 84.33$ N/mm^2; read off

$$f = 145 \text{ N/mm}^2.$$

Then the checks are:

Bottom $p + f'_X = 28.9 + 199.8 = 228.7 \; '< f_y$ $= 255$ N/mm^2

Centre $p' + N_X f_X = 84.38 + 1.06 \times 78.12 = 167.2 > f$ $= 145$ N/mm^2.

An additional restraint is required at the centre of the column. This part of the column could be checked for the unfactored load and moment using elastic design to BS449.

(b) Rafter design

Provide seven purlins spaced at 1750 mm as shown in Fig. 7.18(b). The moments at the purlin points are shown in the figure. Check portion BJ.

Moment ratio	$\beta = 20.27/97.52$	$= 0.21$
Axial stress	$p = 1.88 \times 53.32 \times 10/48.4$	$= 20.71 \text{ N/mm}^2$
From Chart 1/23 [2] the slenderness ratio l/r_{YY}		$= 75$
Allowable spacing of restraints $= 75 \times 28.9$		$= 2168 \text{ mm}$
Spacing provided		$= 1900 \text{ mm}$

Provide restraints to the bottom flange at the purlin at J.

The purlins provide restraints to the compression flange for the hinge near the apex. No stays to the bottom flange are required.

(c) Correction for deflection

$CL/d = 20/(61 \times 0.397)$	$= 0.826$
$L/h_1 = 20/6$	$= 3.33$
Roof slope	$\phi = 22°$
From Chart 4/29 [2] the correction factor	$= 1\%.$
The corrected load factor $= 1.85 - 1\%$	$= 1.83.$

This is satisfactory.

(d) Working load deflection

From Chart 13/29 [2] for $\phi = 22°$ and $L/h_1 = 3.33$

$$\delta/[h_1 L\sigma_y/d\lambda_b]10^6 = 0.64$$

where
$$h_1 = 6\,000 \text{ mm}$$
$$L = 20\,000 \text{ mm}$$
$$d = 397.3 \text{ mm}$$
$$\lambda_b = 1.88$$

$\delta = \dfrac{0.64 \times 6\,000 \times 20\,000 \times 255}{397.\,3 \times 1.88 \times 10^6}$	$= 26.2 \text{ mm}$
$\delta/h_1 = 26.2/6\,000$	$= 1/229.$

(e) Conclusion

The fixed base portal is 13% lighter than pinned base structure. It is also stiffer and the correction to the load factor for deflection is much less. The frame could be redesigned taking a haunch at the eaves into account.

7.6 Coursework exercises

1. Redesign the two pinned steel portal for the warehouse specified in Section 5.7, Exercise 1, using plastic theory.
2. Redesign the frames in Exercise 1 using a fixed base portal.
3. Redesign the two pinned supporting portal for the steel framed building specified in Section 5.7, Exercise 2, using plastic theory.
4. The steel frame for a loading station is shown in Fig. 7.19. The frames are at 6 m centres and the length of the building is 36 m. The roof and walls consist of steel sheeting and insulation supported on purlins and sheeting rails, respectively. The total dead load of the roof and walls is 0.52 kN/m^2. The imposed load on the roof is 0.75 kN/m^2. The building is located on the outskirts of a city.

    Design an intermediate frame using plastic theory.

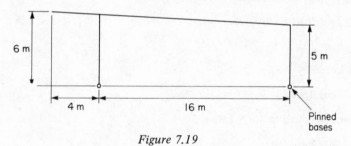

Figure 7.19

References and further reading

[1] Baker, J. F., Horne, M. R. and Heyman, J. (1965). *The Steel Skeleton*, Vol. 11, Plastic Behaviour and Design. Cambridge University Press, Cambridge.
[2] Constrado Plastic Design, Supplement (1979). London.
[3] Horne, M. R. (1971). *Plastic theory of Structures*. Nelson, London.
[4] Horne, M. R. (1964). *The plastic design of columns*. Publication no. 23. B.C.S.A., London.
[5] Horne, M. R. and Chin, M. W. (1966). *Plastic design of portal frames in steel to BS968*. Publication no. 29. B.C.S.A., London.
[6] Morris, L. J. and Randell, A. L. (1975). *Plastic design*. Constrado, London.
[7] Neal, B. G. (1965). *Plastic methods of structural analysis*. Chapman and Hall, London.

[8] *Structural Steelwork Handbook* (1978). B.C.S.A. and Constrado, London.

[9] Horne, M. R. and Ajmani, J. L. (1971). Design of columns restrained by side rails and the post buckling behaviour of laterally restrained columns. *The Structural Engineer*, **49** (8).

[10] Horne, M. R. (1978). *Continuous construction and special problems of single storey construction. The background for the new British Standard for structural steelwork*. Constrado, London.

[11] BS4360 (1972). *Weldable structural steels*. British Standards Institution, London.

8 Roof structures and wide-span buildings

8.1 General considerations

Roof structures and wide-span buildings covering large areas without intermediate columns are required for supermarkets, exhibition buildings, warehouses, factories, sports halls and stadia, swimming pools, etc. A general classification is as follows:

(a) roof structures such as long-span lattice girders or space decks supported on perimeter columns;
(b) wide-span building frames such as portals or arches;
(c) structural forms such as barrel vaults, domes or hyperbolic paraboloids.

It is convenient to classify these structures according to the methods that are required for their analysis and design. This gives:

(a) statically determinate systems where the simple design method can be used. These include, beams, trusses, lattice girders, three pinned portals or arches and braced ribbed domes. Essentially these systems are divided into sets of one-way spanning elements for design;
(b) statically indeterminate systems where elastic rigid design is used. These include two pinned and fixed base portals and arches which are one-way spanning structures and the two-way spanning space structures such as space decks, barrel vaults and domes.

One way spanning structures are considered in this chapter. Space structures are considered in Chapter 9.

Other solutions are also used such as stressed skin and folded plate structures and tension structures in the form of either nets or cable girders. These types of structures are not considered in this book.

8.2 Traditional simple design

Simple design is used with systems of beams, trusses and lattice girders. The one-way spanning elements may be in single or double systems. Some commonly used systems and elements are discussed below. Horizontal wind bracing is required with roofs of sheeting or decking to carry wind loads to the wall bracing. Cast *in situ* or tied precast concrete slab roofs will act as horizontal diaphragms in transmitting loads.

8.2.1 Single and double beam or lattice girder systems

Two systems for a flat roof structure one with primary beams only spanning the full width of the building and the other with primary lattice girders and secondary purlins supporting the roof deck are shown in Fig. 8.1(a) and (b), respectively. It is necessary to make comparative designs to see which system gives the most economical solution.

Castellated beams are used for long spans carrying relatively light loading. Parallel chord lattice girders fabricated from structural hollow sections are also available.

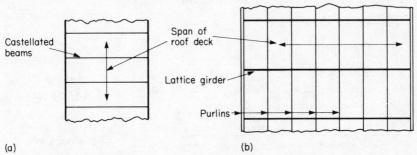

Figure 8.1 (a) Castellated beams spanning one way. (b) Primary lattice girders and secondary units.

8.2.2 Example – comparative design

(a) Specification

Make comparative designs for a flat roof of 20 m span. The roof dead load including asphalt, screed, precast units, steel and ceiling is 3.8 kN/m^2. The imposed load is 1.5 kN/m^2. The precast units span 4.0 m. Use Grade 50 steel.

(b) First scheme – castellated beams at 4.0 m centres

The arrangement for the roof is similar to that shown in Fig. 8.1(a)

Load per beam $= (3.8 + 1.5)4 \times 20$		$= 424$ kN
Moment $= 424 \times 20/8$		$= 1060$ kN
Modulus $= 1060 \times 10^3/230$		$= 4609$ cm^3

Require a castellated UB 1029 x 254 x 125 kg/m,

where $Z = 5375$ cm^3.

Weight of the support steel $= 125/4$ $= 31.2$ kg/m^2.

(c) Second scheme — lattice girders at 4.0 m centres

The arrangement is the same as for the first scheme except that lattice girders 1.5 m deep will be used instead of the castellated beams. The distributed load = $(3.8 + 1.5)4 = 21.2$ kN/m. The arrangement of the truss is shown in Fig. 8.2(a). The truss may be analysed by joint resolution. The forces in the members are also shown in the figure. The fixed end moment in the top chord

$$= 21.2 \times 2^2/12 \qquad\qquad = 7.07 \text{ kN m.}$$

Top chord C D E F

Compression = 692.5 kN

Moment   =   7.07 kN m.

Try 150 x 150 x 8 RHS.

$A = 45.1 \text{ cm}^2, Z = 201 \text{ cm}^3, r = 5.78 \text{ cm}$

$l/r = 0.85 \times 2000/57.8 \qquad\qquad = 29.4$

$p_c = 200 \text{ N/mm}^2$

$p_{bc} = 230 \text{ N/mm}^2$

$f_c = 692.5 \times 10/45.1 \qquad\qquad = 153.5 \text{ N/mm}^2$

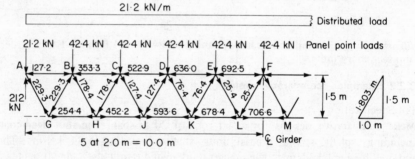

(a)

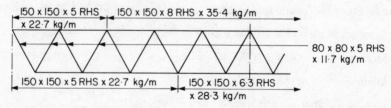

All other internal members except those noted : 70 x 70 x 3·6 RHS x 7·48 kg/m

(b)

Figure 8.2   (a) Truss arrangement and member forces (kN). (b) Truss members.

$f_{bc} = 7.07 \times 10^3/201$ $= 35.2 \text{ N/mm}^2$

$f_c/p_c + f_{bc}/p_{bc} = 0.768 + 0.153$ $= 0.921$

Use 150 x 150 x 8 RHS x 35.4 kg/m.

Top chord A B C
Compression = 353.3 kN

Moment = 7.07 kN m

Try 150 x 150 x 5 RHS.

$A = 28.9 \text{ cm}^2, Z = 157 \text{ cm}^3, r = 5.91 \text{ cm}$

$l/r = 0.85 \times 2000/59.1$ $= 28.8$

$p_c = 200 \text{ N/mm}^2$

$f_c = 353.3 \times 10/28.9$ $= 122.2 \text{ N/mm}^2$

$f_{bc} = 7.07 \times 10^3/157$ $= 45.0 \text{ N/mm}^2$

$f_c/p_c + f_{bc}/p_{bc} = 0.611 + 0.196$ $= 0.807$

Use 150 x 150 x 5 RHS x 22.7 kg/m.

Bottom chord K L M
Tension = 706.6 kN

$A = 706.7 \times 10/215$ $= 32.86 \text{ cm}^2$

Use 150 x 150 x 6.3 RHS x 28.3 kg/m: $A = 36 \text{ cm}^2$.

Bottom chord N G H J K
Tension = 593.6 kN

$A = 593.6 \times 10/215$ $= 27.6 \text{ cm}^2$.

Use 150 x 150 x 5 RHS x 22.7 kg/m $A = 28.9 \text{ cm}^2$

Internal members AG, GB, BH, HC
Compression = 229.3 kN

Try 80 x 80 x 5 RHS.

$A = 14.9 \text{ cm}^2, r = 3.05 \text{ cm}$

$l/r = 1802 \times 0.85/30.5$ $= 50.2$

$p_c = 184 \text{ N/mm}^2$

$f_c = 229.3 \times 10/14.9$ $= 153.8 \text{ N/mm}^2$

Use 80 x 80 x 5 RHS x 11.7 kg/m.

Internal members CJ, JD, DK, KE, EJ, LF
  Compression = 127.4 kN

  Try 70 x 70 x 3.6 RHS.

  A = 9.5 cm^2, r = 2.7 cm

  l/r = 1802 x 0.85/27   = 56.7

  p_c = 175 N/mm^2

  f_c = 127.4 x 10/9.5   = 134.1 N/mm^2

  Use 70 x 70 x 3.6 RHS x 7.48 kg/m.

  The truss members are shown in Fig. 8.2(b).

  Weight = (12 x 35.4) + (8 x 22.7) + (6 x 28.3) + (14 x 22.7)

  　　　　　 + (8 x 1.8 x 11.7) + (12 x 1.8 x 7.48)

  　　　　 = 1424.1 kg

  Weight/m^2 = 1424.1/(20 x 4)   = 17.8 kg/m^2.

The lattice girder can be made lighter by increasing the depth. Buckling in the web members then becomes more important. The increase in the building height must also be considered.

(d) Third scheme — Primary girders at 8 m centres and secondary beams at 4 m centres

The arrangement of the roof steel is shown in Fig. 8.3. The precast slabs span

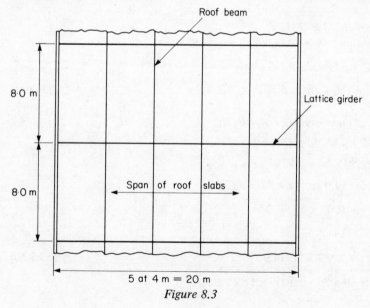

Figure 8.3

across the width of the building in this case.

Roof beam

Load	$= (3.8 + 1.5)4 \times 8$	$= 169.6$ kN
Moment	$= 169.6 \times 8/8$	$= 169.6$ kN m
Modulus Z	$= 169.6 \times 10^3/230$	$= 737.4$ cm^3

406×140 UB 46 $\quad Z = 777.8$ cm^3

Weight of four beams $= 32 \times 46$ $\qquad\qquad\qquad$ $= 1472$ kg

Adopt a lattice girder 2 m deep as shown in Fig. 8.4(a). Assume the self weight is 2 kN/m. The panel point loads are shown on the figure. The truss is analysed by joint resolution and the member forces are shown on the diagram of the truss.

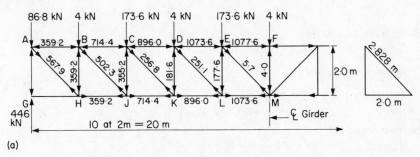

(a)

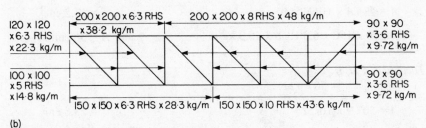

(b)

Figure 8.4 Lattice girder. (a) Truss arrangement, loads and member forces (kN). (b) Truss members.

Top chord C D E F

Compression $= 1077.6$ kN

Try $200 \times 200 \times 8$ RHS.

$A = 61.1$ cm^2, $r = 7.83$ cm

$l/r = 4000/78.3$ $\qquad\qquad\qquad\qquad\qquad$ $= 51.1$

$p_c = 182.8$ N/mm^2

$$f_c = 1077.6 \times 10/61.1 \qquad\qquad = 176.4 \text{ N/mm}^2$$

Use 200 x 200 x 8 RHS x 48 kg/m.

Top chord A B C
Compression = 714.4 kN

Try 200 x 200 x 6.3 RHS.

$A = 48.6 \text{ cm}^2$, $r = 7.9$ cm

$l/r = 4000/79 \qquad\qquad = 50.6$

$p_c = 183 \text{ N/mm}^2$

$f_c = 714.4 \times 10/48.6 \qquad\qquad = 146.9 \text{ N/mm}^2$

Use 200 x 200 x 6.3 RHS x 38.2 kg/m.

Bottom chord K L M
Tension   = 1073.6 kN

$A = 1073.6 \times 10/215 \qquad\qquad = 49.9 \text{ cm}^2$

Use 150 x 150 x 10 RHS x 43.6 kg/m: $A = 55.5 \text{ cm}^2$

Bottom chord G H J K
Tension   = 714.4 kN

$A = 714.4 \times 10/215 \qquad\qquad = 33.2 \text{ cm}^2$

Use 150 x 150 x 6.3 RHS x 28.3 kg/m: $A = 36 \text{ cm}^2$.

Diagonals A H, B J
Tension   = 507.9 kN

$A = 507.9 \times 10/215 \qquad\qquad = 23.6 \text{ cm}^2$

Use 120 x 120 x 6.3 RHS x 22.3 kg/m: $A = 28.5 \text{ cm}^2$.

Diagonals C K, D L, E M
Tension   = 256.8 kN

$A = 256.8 \times 10/215 \qquad\qquad = 11.9 \text{ cm}^2$

Use 90 x 90 x 3.6 RHS x 9.72 kg/m: $A = 12.4 \text{ cm}^2$.

Verticals B H, C J
Compression = 359.2 kN

Try 100 x 100 x 5.0 RHS.

$A = 18.9$ cm^2, $r = 3.87$ cm

$l/r = 0.85 \times 2000/38.7$ $\hspace{4cm}$ $= 43.9$

$p_c = 191$ N/mm^2

$f_c = 359.2 \times 10/18.9$ $\hspace{4cm}$ $= 190$ N/mm^2

Use 100 x 100 x 5.0 RHS x 14.8 kg/m.

Verticals D K, E L, F M
Compression = 181.6 kN

Try 90 x 90 x 36 RHS.

$A = 12.4$ cm^2, $r = 3.52$ cm

$l/r = 0.85 \times 2000/35.2$ $\hspace{4cm}$ $= 48.3$

$p_c = 185$ N/mm^2

$f_c = 181.6 \times 10/12.4$ $\hspace{4cm}$ $= 145$ N/mm^2

Use 90 x 90 x 3.6 RHS x 9.72 kg/m.

The truss members are shown in Fig. 8.4(b).

Weight = (12 x 48) + (8 x 38.2) + (8 x 43.6) + (12 x 38.3)

$\hspace{2cm}$ + (8 x 14.8) + (10 x 9.72) + (4 x 2.828 x 22.3)

$\hspace{2cm}$ + (6 x 2.828 x 9.72)

$\hspace{2cm}$ = 2207.9 kg

Total weight of roof steel = 2207.8 + 1472 $\hspace{2cm}$ = 3679.8 kg

Weight/m^2 = 3679.8/8 x 20 $\hspace{3cm}$ = 22.9 kg/m^2

The second scheme is the most economical in this case.

8.3 Sawtooth roof construction

This was a very common type of construction for older warehouses and factories where a clear internal floor area was required. The sawtooth roof permits glazing to be used to admit natural lighting to the interior of the building. The framing for a typical building is shown in Fig. 8.5. Other variations of this type of construction are possible where the main lattice girder spans in the same direction as the roof trusses.

The analysis and design of the structure is set out briefly below. See Fig. 8.5.

(a) Purlins carry roof sheeting and insulation. Glazing support steel is provided as noted on Section AA.

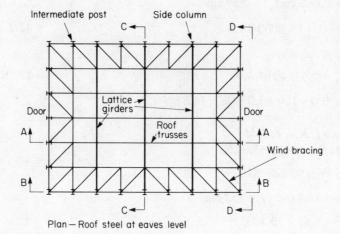

Plan — Roof steel at eaves level

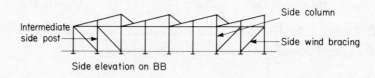

Part plan — Bracing in top chord of roof trusses — End bays

Section AA

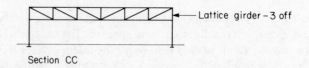

Side elevation on BB

Section CC

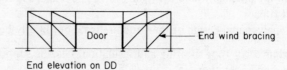

End elevation on DD

Figure 8.5

(b) Triangular roof trusses frame between the lattice girders and are designed to carry roof loading. See Section AA.

(c) Lattice girders span the full width of the building and support the roof truss loads from (b). See Section CC.

(d) The main side columns are designed to carry the lattice girder reactions. Intermediate posts are provided on the sides between the main columns. Columns on the ends carry roof truss loads.

(e) Bracing is required in the end bays of the sloping top chords of the roof truss to carry wind load on the side of the roof truss and wind drag on the roof. This also provides lateral stability for the top chords of the trusses. Refer to Part plan – Bracing in the top chord.

(f) Wind trusses are provided in both directions at eaves level to carry wind loads on the sides and roof of the building. Refer to Plan of roof steel.

(g) Wind bracing is provided on the ends and sides of the building in the end bays. This carries the horizontal loads to the foundations.

The design of this type of building is set as a coursework problem.

8.4 Arch roofs

8.4.1 Types and construction

Arch or barrel vault roofs are constructed generally in the form of circular arcs, but parabolic arches may also be used. This structural form is used for roofing sports areas such as gymnasiums and swimming pools or for various types of industrial buildings, such as, train and bus sheds, market buildings, warehouses, etc. The Victorian railway stations have many fine examples of multiple steel arch roofs.

Some of the types of arch roofs are shown in Fig. 8.6(a), (b) and (c). Some general comments on the construction are given below.

(a) The arch frames may form the whole building with the abutments set at ground level as shown in Fig. 8.6(a). The roof is carried by the arch rings which are analysed as plane frames. The arch may be three pinned, two pinned or fixed as shown in Fig. 8.6(d). The arch rib may be composed of a single member as shown in Fig. 8.6. Alternatively, braced construction can be used as shown in Fig. 8.10(b). Bracing is provided in the end bays to stabilize the buildings and resist longitudinal wind load.

(b) Barrel vault roofs supported on abutments are shown in Fig. 8.6(b) and (c). A braced barrel vault is shown in Fig. 8.6(b) and a three-way grid is shown in Fig. 8.6(c). Other types of framing are also used. In the three-way grid the loading is carried on the four walls of the building. These structures are more rigid and stresses due to unsymmetrical loading are distributed more evenly throughout the structure than is the case in the arch roof consisting of arch ribs discussed in (a). These structures approximate to cylindrical shell roofs. These frameworks should be analysed as space frames. However, in the

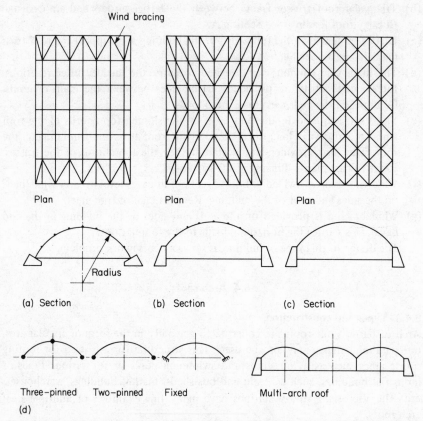

Figure 8.6   (a) Arch roof building. (b) Braced barrel vault roof. (c) Three-way grid barrel vault roof. (d) Types of arch.

past it has been customary to treat a bay or unit length as an equivalent plane frame for analysis. These structures will not be discussed further here.

8.4.2 Loading, analysis and design

(a) Loading
The loading is due to dead, imposed and wind loads. The arch must be analysed for cases where the imposed loading covers part of the roof as this will result in more severe bending moments than when the load is applied to the whole roof. Wind loading must also be considered and data on wind loading is given in [6].

(b) Analysis
A major design decision is how the horizontal thrust is to be resisted. Two methods are used. These are:

(i) the thrust is carried on the foundations either directly at foundation level if the arch ring is carried to ground level or on buttressed walls or piers where the arch provides the roof only. See Fig. 8.7(a). Some space is lost when the arch abutments are at ground level, but the foundations are more economic;

(ii) the thrust is taken by ties between the arch abutments. The foundations then carry primarily axial load but must also resist overturning due to wind loads. The ties are usually steel bars. The dead load must be sufficient to ensure the uplift due to wind will not cause a reversal of force in the ties. See Fig. 8.7(b).

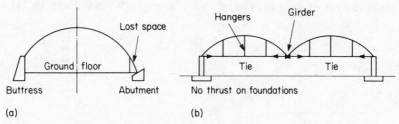

(a)   (b)

Figure 8.7 (a) Thrust taken on foundations. (b) Tied arches.

The three-pinned arch is statically determinate and the analysis may be carried out manually. For an arch composed of single member ribs see Fig. 8.6(d), the reactions at the pins are found first and then the moment thrusts and shears can be found at all points by statics. If the arch is subjected to uniform vertical loading a parabolic arch will be in pure compression while a circular arch will be in bending as well as compression. Asymmetrical loading causes bending in all arches. For the braced pin jointed three-pinned arch shown in Fig. 8.10(a) the reactions at the pins are determined first. Then the forces on the members may be found by joint resolution or by a force diagram. This frame may also be classed as a three-pinned braced portal.

The fixed arch and two-pinned arch are statically indeterminate. The analysis may be carried out manually using the strain energy or virtual work method and the curved arch members may be taken into account. Alternatively, a standard plane frame computer program may be used where the arch is divided by joints into a given number of sections. The arch members are assumed to be straight between the joints. The braced arch shown in Fig. 8.8(b) can be analysed using the plane frame program with pinned or fixed joints. Various cases with the imposed load covering part of the roof can be examined to determine the worst case for design.

(c) Design

The arch rib is designed for thrust and bending moment. The arch may be constructed from a single section such as a univeral beam or structural hollow section. The stability of the arch must be given careful consideration because buckling can occur both in the plane of the arch and perpendicular to it. This is set out briefly here.

(i) Stability in the plane of the arch [4, 5].

The buckling modes for the three-pinned, two-pinned and fixed arch are shown in Fig. 8.8. The three-pinned arch buckles by the crown moving down. The two-pinned and fixed arch buckle into an antisymmetrical shape with a point of contraflexure near the crown. It can be shown theoretically that the arch can be treated as a straight column of length S equal to half the arch length. The effective length depends on the end conditions and the span and rise of the arch. The following factors are given in Johnston [4].

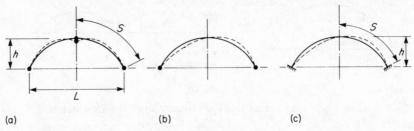

(a)   (b)   (c)

Figure 8.8   (a) Three-pinned arch. (b) Two-pinned arch. (c) Fixed arch.

Fixed arch — the half length is pinned at one end fixed at the other

$l = (0.7 \text{ to } 0.77) \times S$.

Two-pinned and three-pinned arch — the half length is pinned at both ends

$l = (1.02 \text{ to } 1.25) \times S$.

The higher effective length factors apply to high-rise arches. Johnston [4] and Timoshenko and Gere [5] should be consulted for further information. Special care must be taken with slender arches where the deformations must be taken into account in calculating moments.

(ii) Lateral stability

The arch requires lateral support to prevent both overall buckling and also lateral instability of the lower flange when this is in compression due to bending and axial load. The lower flange of a universal beam requires support at specified intervals along its length, in respect to bending. Structural hollow sections will be stable provided the depth does not exceed four times the width. See Clause 19(d) of BS449 [1]. However, it is considered that the whole section requires supporting as the purlins attached to the top flange only cannot be relied on to prevent overall buckling, particularly in portions AB and DE, Fig. 8.9, where the lower flange is in compression due to bending from dead and imposed loads. Uplift from wind load can cause reversal of stress in the centre section and a support may be needed at the

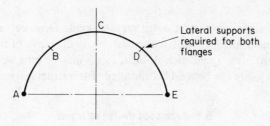

Figure 8.9

centre C. Possible lateral support points for the arch rib are shown in the figure. Heavy stiff roof construction can be assumed to give adequate lateral support to shallow depth arch ribs where the decking is attached directly to the ribs.

With regard to bending of structural hollow sections it should be noted that some sections with thin walls can suffer local instability before reaching the full elastic moments. These sections and their permissible compressive stresses in bending are listed in the Constrado *Structural Steelwork Handbook* [3] for structural hollow sections.

Braced construction can also be used for the arch as shown in Fig. 8.10. The lower chord of the arch must be supported at intervals to prevent buckling. The support may be provided by longitudinal latticed members or by bracing from the purlins.

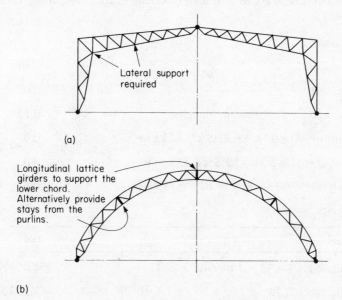

Figure 8.10 (a) Three-pinned braced portal. (b) Two-pinned braced arch.

(d) Construction

For arches of single members the rib may be made from curved members or from short straight members where the joints lie on the curve of the arch. Rolled sections are bent after application of heat. Joints may be butt welded or bolted. Flanged bolted joints can be used for ribs made from structural hollow sections.

8.5 Arch roof design example

8.5.1 Specification

A sports building is to be constructed using two-pinned steel arches. The building has brick side walls 2.5 m high and the span of the enclosed portion is 20 m. The span of the arch is 24.4 m between abutments and the rise is 6.5 m. The arches are spaced at 4.5 m and the length of the building is 36 m. A section through the building is shown in Fig. 8.11. The end bays are braced to resist longitudinal wind load; refer Fig. 8.6(a). The cladding will be steel sheet on insulation supported on purlins of structural hollow section. Design the arch rib using a rectangular hollow section in Grade 43 steel.

8.5.2 Arrangement for analysis and joint co-ordinates

The roof part of the arch will be divided into 12 segments to give a purlin spacing of 1.84 m. The arch will be assumed to be connected by straight members between joints located at the abutments and the purlin points. Loads will be applied at the purlin points. The joint co-ordinates are shown in Fig. 8.13(a).

8.5.3 Loading

(a) Dead load	kN/m^2
Sheeting	0.09
Insulation — plasterboard, fibreglass	0.12
Purlins — 90 x 90 x 3.6 RHS x 9.72 kg/m	0.07
Bracing — 50 x 50 x 4 RHS x 5.72 kg/m	0.02
Arch rib — 300 x 200 x 10 RHS x 75 kg/m	0.17
Fixings, services	0.02
Total	0.49

Joint loads 3 − 13 = 0.49 x 4.5 x 1.84   = 4.06 kN

Joint loads 2, 14   = 2.03 + 1.73 x 4.5 x 0.19   = 3.51 kN

The dead loads are shown in Fig. 8.11(b).

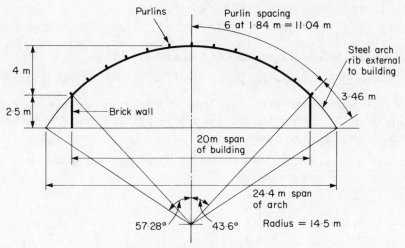

Figure 8.11 Section through the building.

(b) Imposed load No access. Load on plan   = 0.75 kN/m^2

Joint loads 2, 14   = (10 − 8.59)0.75 x 4.5/2   = 2.39 kN

Joint loads 3, 13   = (10 − 7.04)0.75 x 4.5/2   = 4.99 kN

Joint loads 4, 12   = (8.59 − 5.38)0.75 x 4.5/2   = 5.42 kN

Joint loads 5, 11   = (7.04 − 3.64)0.75 x 4.5/2   = 5.74 kN

Joint loads 6, 10   = (5.38 − 1.83)0.75 x 4.5/2   = 5.99 kN

Joint loads 7, 9   = 3.64 x 0.75 x 4.5/2   = 6.14 kN

Joint load 8   = 1.83 x 0.75 x 4.5   = 6.18 kN

The arch will be analysed for cases where the imposed load covers

(i) the whole arch,
(ii) 60% of the arch. It can be established by trials that this case gives the maximum moment.

The imposed loads are shown in Fig. 8.13(c).

(c) Wind loads

Basic wind speed V = 45 m/s

Ground roughness — category 3

Building size — class B

Height = 6.5 m

Factor S_2 from Table 3 of BS449 = 0.68

Factor S_1, S_3 = 1.0

Design wind speed V_s = 0.68 x 45 = 30.6 m/s

Dynamic pressure q = 0.613 x $30.6^2/10^3$ = 0.574 kN/m^2

External pressure coefficients C_{pe}. These are taken from Newberry and Eaton [6], Table 11.5, for a rectangular clad building with an arch roof. The external pressure coefficients are shown on Fig. 8.12.

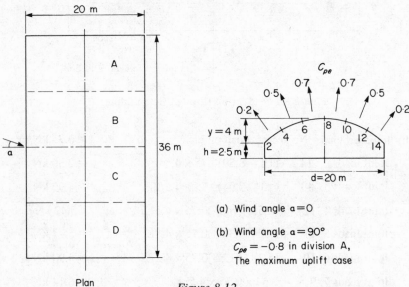

Plan

Figure 8.12

The internal pressure coefficient C_{pi} is taken as the more onerous of −0.2 or −0.3. The internal pressure case only will be considered here as this gives the maximum uplift.

Note that the wind loads on the brick walls do not introduce load into the arch ring.

The wind loads for the two cases where $\alpha = 0°$ and $\alpha = 90°$ are calculated in Table 8.1. For entry of loads in the computer program, the vertical and horizontal components are calculated.

8.5.4 Computer analysis

(a) Program input

The section of the arch rib assumed for analysis is a 300 x 200 x 10 RHS. The properties are:

A = 95.5 cm^2

I = 11 940 cm^4.

Table 8.1
Wind: Loads kN   Wind angle α = 0°   Internal pressure case

Joint	Normal load (kN)			Angle to vertical	Vertical component	Horizontal component
2	0.7	x 0.574 x 0.92 x 4.5	1.66	43.6°	+1.2	−1.15
3	0.7	x 0.574 x 1.84 x 4.5	3.33	36.3°	+2.68	−1.97
4		"	3.33	29.1°	+2.91	−1.62
5		"	3.33	21.8°	+3.09	−1.24
6	0.8	x 0.574 x 1.84 x 4.5	3.8	14.5°	+3.67	−0.95
7	0.9	x 0.574 x 1.84 x 4.5	4.28	7.3°	+4.26	−0.54
8		"	4.28	0	+4.28	
9		"	4.28	7.3°	+4.26	+0.54
10	0.8	x 0.574 x 1.84 x 4.5	3.8	14.5°	+3.67	+0.95
11	0.7	x 0.574 x 1.84 x 4.5	3.33	21.8°	+3.09	+1.24
12	0.55	x 0.574 x 1.84 x 4.5	2.61	29.1°	+2.28	+1.27
13	0.4	x 0.574 x 1.84 x 4.5	1.9	36.3°	+1.53	+1.12
14	0.4	x 0.574 x 0.92 x 4.5	0.95	43.6°	+0.69	+0.66

Wind loads kN   Wind angle α = 0   Internal pressure case

Joint	Normal load (kN)		Angle to vertical	Vertical component	Horizontal component
2	0.574 x 0.92 x 4.5	2.38	43.6°	1.72	−1.64
3	0.574 x 1.84 x 4.5	4.75	36.3°	3.83	−2.81
4	"	4.75	29.1°	4.15	−2.31
5	"	4.75	21.8°	4.41	−1.76
6	"	4.75	14.5°	4.6	−1.19
7	"	4.75	7.3°	4.71	−0.6
8	"	4.75	0	4.75	0

The input for the plane frame program is taken from Sections 8.5.2 and 8.5.3 above. The joints are located at the divisions shown in Fig. 8.13(a). The arch is analysed for the following load cases:

dead load;
imposed load over the whole arch;
imposed load over 60% of the arch;
wind load, internal pressure, angle α = 0°;
wind load, internal pressure, angle α = 90°.

(b) Program output
The results of the computer analysis for the five load cases are shown in Fig. 8.14. Here the diagrams show

(i) the vertical and horizontal reactions at the abutments,
(ii) the axial compression or tension in all arch members,
(iii) the bending moment diagram giving the moments at the joints of the arch.

Note that the shear forces in the arch rib are very small and have not been listed in the figure.

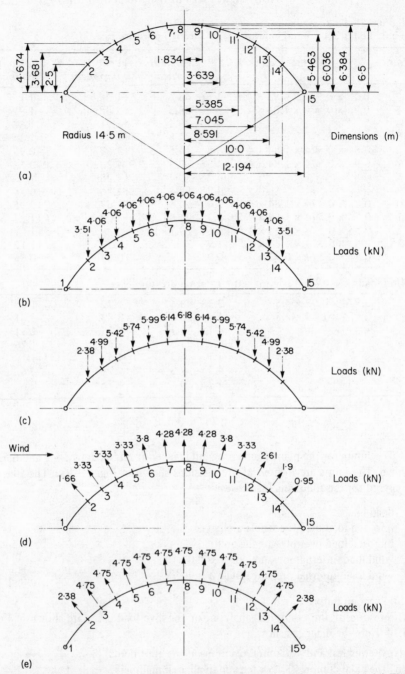

Figure 8.13 (a) Joint co-ordinates. (b) Dead loads. (c) Imposed loads. (d) Wind loads – internal pressure case – wind angle $\alpha = 0^\circ$. (e) Wind loads – internal pressure case – angle $\alpha = 90^\circ$.

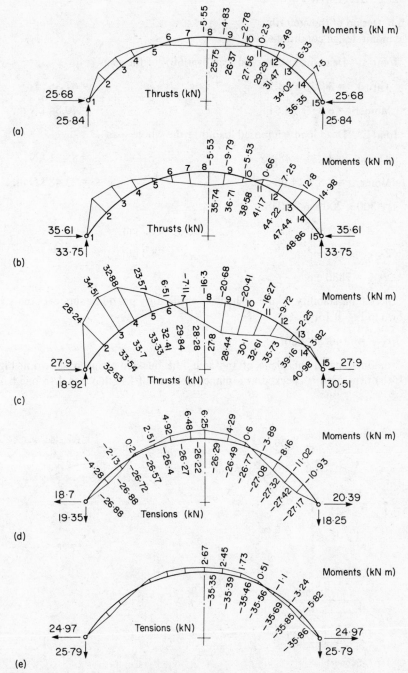

Figure 8.14 (a) Dead load. (b) Imposed load over whole span. (c) Imposed load over 60% of span, joints 6–14. (d) Wind load – internal pressure case – angle $\alpha = 0°$. (e) Wind load – internal pressure case – angle $\alpha = 90°$.

8.5.5 Design of the arch rib

Maximum design conditions:

Joint 3 — Dead load + imposed load over 60% of the span

Thrust	= 34.02 + 33.54	= 67.56 kN
Moment	= 6.33 + 34.51	= 40.84 kN m.

Joint 2 — Dead load + imposed load over the whole span

Thrust	= 36.35 + 48.86	= 85.21 kN
Moment	= 7.5 + 14.98	= 22.48 kN m.

Try 300 x 200 x 6.3 RHS.

$$A = 61.2 \text{ cm}^2, \qquad r_{XX} = 11.3 \text{ cm}$$

$$Z_{XX} = 525 \text{ cm}^3, \qquad r_{YY} = 8.3 \text{ cm}$$

$$I_{XX} = 7880 \text{ cm}^4.$$

The lateral stability of the arch rib is ensured by providing support points as shown in Fig. 8.15(a)

$$l/r_{YY} = 7140/83 = 86.$$

For buckling in the plane of the arch, the buckled shape is shown in Fig. 8.15(b). The arch is checked as a pinned column of length S equal to one half

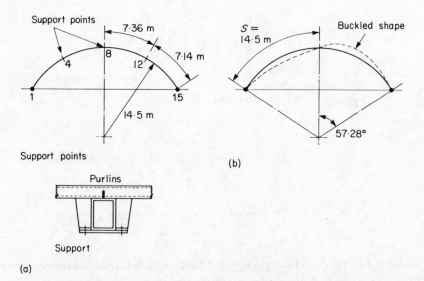

Figure 8.15   (a) Lateral support for arch. (b) In-plane buckling of arch.

the length of rib. The effective length factor from Johnston [4] is 1.1.

$l/r_{XX} = 1.1 \times 14\,500/113 = 141.2$

$p_c = 45\ \text{N/mm}^2$ from Table 17a of BS449.

The permissible bending stress is taken from the Constrado *Structural Steel-work Handbook* [3] for structural hollow sections. The trial section is subject to local instability at the full elastic moment.

$p_{bc} = 151\ \text{N/mm}^2$

The actual stresses are:

$f_c = 67.56 \times 10/61.2$ $= 11.04\ \text{N/mm}^2$

$f_{bc} = 40.84 \times 10^3/525$ $= 77.81\ \text{N/mm}^2$

$$\frac{f_c}{p_c} + \frac{f_{bc}}{p_{bc}} = \frac{11.04}{45} + \frac{77.81}{151} = 0.25 + 0.52 = 0.77.$$

The next lighter section 200 x 150 x 6.3 RHS is overstressed. Adopt 300 x 200 x 6.3 RHS.

8.5.6 Construction
The arches can be fabricated in two parts for transport to site. The structural hollow section can be bent to the radius required after heating. The centre site joint can be made by site welding. The pinned base consists of a plate welded to the arch rib with two holding-down bolts.

8.6 Coursework exercises

1. A warehouse is 60 m long x 40 m wide x 6 m clear height to underside of roof. The interior must be clear of all columns. Draw framing plans for a structure with a sawtooth roof. Design the roof trusses, lattice girders and supporting columns.
2. A large conference/exhibition hall 72 m long is to be roofed with an arch structure of 50 m span between abutments set at ground level and 10 m rise above floor level. The arches are to be at 6 m centres. Set out a circular three-pinned braced arch with depth of arch between inside and outside chords of 1.0 m. Draw the framing plans for the structure. Analyse and design the arch rib.

References and further reading

[1] BS449, Part 2 (1969). *The use of Structural Steel in Building*. British Standards Institution, London.
[2] *Handbook on Structural Steelwork* (1978). B.C.S.A. and Constrado, London.

[3] *Structural Steelwork Handbook* (1976). Structural hollow sections to BS4848, Part 2, Constrado, London.

[4] Johnston, B. G. (Ed.) (1976). *Guide to stability design criteria for metal structures*, 3rd ed. John Wiley, New York.

[5] Timoshenko, S. P. and Gere, J. M. (1961). *Theory of elastic stability*, 2nd ed. McGraw-Hill, New York.

[6] Newberry, C. W. and Eaton, K. J. (1974). *Wind loading handbook*. London, Building Research Establishment, H.M.S.O., London.

9 Space structures

9.1 General considerations

This chapter is again concerned with roof structures covering large areas. Here the structures considered are

(a) two-way spanning roof systems – grids and space decks;
(b) domes which may form the roof only or the complete structure.

The space deck provides a rigid flat roof structure capable of spanning large distances with a small depth of construction. The small depth reduces the building height and cost of cladding and also saves internal space requiring heating. The exposed space deck forms a pleasing structure which is very popular at the present time. The three-dimensional form ensures that resistance to concentrated and asymmetrical roof loads is supplied by many members thus reducing maximum member loads and roof deflections.

The large framed dome is one of the most spectacular and pleasing of structures. Domes are used to cover sports arenas, auditoria, exhibition areas, churches, etc. The dome in shell form is an ancient type of structure.

In the previous chapter, the construction of the roof structures was such that they could be idealized into two-dimensional structures for analysis and design which can then be assembled to form the roof. Here, the complete two-dimensional grid or three-dimensional space deck or dome is analysed. These structures are highly redundant. Thrust, bending about both axes and torsion are considered in a general case.

9.2 Space grids

9.2.1 Two-way spanning roofs

If the length of the area to be roofed is more than twice the breadth it is more economical to span one way using the type of construction discussed in Chapter 8. If the area is square or where the length is less than twice the breadth the more economical solution, theoretically, is to span two ways. A rectangular area may be divided into square or near square areas with lattice girders and then two-way spanning structures can be used on the areas of the subdivided roof. See Fig. 9.1.

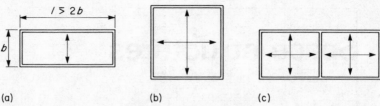

Figure 9.1 (a) One-way spanning structure. (b) Two-way spanning structure. (c) Roof subdivided for two-way spanning structure.

A two-way spanning flat roof structure is termed a grid. These may be single or double layer. The single layer or planar grid consists of an intersecting system of beams. The joints are rigid and torsion is taken into account in the analysis. The analysis can be made using a grid or a general space frame program. The joints are costly to make, distortion is a problem and such structures are not generally used. A planar grid is shown in Fig. 9.2. The application of the load depends on the type of deck or roof unit used. Load from a two-way slab or roof unit is shown in the figure.

The double layer grid may be constructed in a number of ways. Lattice girders or Vierendeel girders intersecting at right angles form a two-way double layer grid. The lower chord lies below the upper chord. The roof is divided into squares which can be covered with roof units such as translucent plastic pyramids. The double layer grid of this type is shown in Fig. 9.3. Three-way grids of similar construction where the roof surface is divided into equilateral triangles are shown in Fig. 9.3(b).

The case where the bottom chord does not lie in the same vertical plane as the top chord or in some cases does not have the same geometrical pattern is known as the space grid or space deck. This is discussed in more detail in the next section.

9.2.2 Space grids — types and construction

(a) Description
Full information on the arrangement, analysis and construction of space grids is given in the British Steel Corporation, Tube Division, publications on space

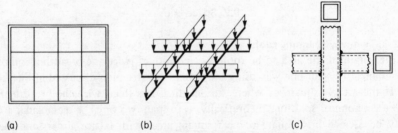

Figure 9.2 (a) Grid. (b) Loads distributed along members. (c) Welded joint for grid in RHS.

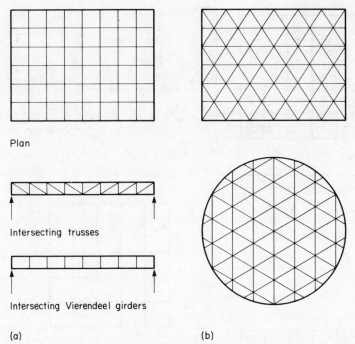

Plan

Intersecting trusses

Intersecting Vierendeel girders

(a)   (b)

Figure 9.3   (a) Two-way double layer grid. (b) Three-way double layer grids.

frame grids [1]. The classification set out in these publications is used here. This information is given here with kind permission of the British Steel Corporation.

The commonest form of two-way grid is shown in Fig. 9.4(a). This is termed the square on square offset with cornice edge. The top and bottom chords form a series of squares. The basic unit is the inverted square based pyramid. One possible variation in the lower chord pattern is shown in Fig. 9.4(b). This pattern is termed square on larger square set diagonally with cornice edges. Other variations are possible.

A great deal of effort in inventiveness, research and testing has been put into perfecting the systems used for constructing space decks. Construction methods may be classified into the following two types:

(i) division of the grid into basic pyramid units;
(ii) joint systems connecting straight grid members.

These systems are discussed in detail below.

(b) Basic pyramid units

For the space grid shown in Fig. 9.4(a) the basic unit is the inverted pyramid. The pyramid units are shop-welded fabrications. The top chords are usually made from angles or channels, the web members from tubular sections and the

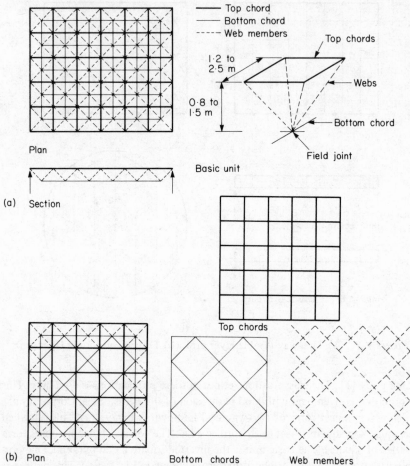

Figure 9.4   (a) Square on square offset with cornice edge. (b) Square on larger square set diagonally — cornice edge.

bottom chords from high-tensile bars with screwed ends. The length of the top chord on the unit varies from about 1.2 m to 2.5 m while depths vary from 0.8 m to 1.5 m. The pyramids conveniently stack together apex upwards for transport to site.

The grid is assembled by bolting the top chords of the pyramids together and connecting the bottom chords through the screwed joints. The whole grid is constructed at ground level and hoisted into position as a complete structure. The Space Deck and Unibat systems are based on the unit pyramid.

Two other systems based on either a triangular or hexagonal subdivision of the top and bottom gird surfaces have also been used. Division into equilateral triangles gives a three-way grid. These grids can be broken down into basic units for ease of construction. Triangular trusses have also been used as the shop-fabricated unit in one system.

(c) The joint system

The joint system gives far greater flexibility in space grid construction than the fixed component system. The difficulty, however, lies in producing an efficient joint which must be capable of connecting eight members coming from various directions in space to meet at a point. The finished structure is assembled from straight members, usually structural hollow sections and joints. Two of the main types used here are the Nodus and Mero systems.

The Nodus joint, illustrated in Fig. 9.5, was developed by the British Steel Corporation Tubes Division. It consists of two half castings clamped together by a high-strength bolt. Four connectors for web members are cast integral with the inner half of the joint. The chord members have grooved ends to lock them into grooves in the joint casting. The web members have forked end connections which are pinned to lugs on the joint casting. Depending on the angle of web members, their centre lines may not meet at the intersection point of the chord members. Thus unequal loads in web members introduce eccentricities and moments in the chord members. The joint is equal to the strength of the members

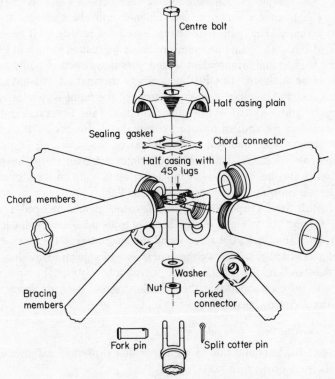

Figure 9.5   The Nodus joint.

it joins and for the top and bottom chord members it is able to take the full fixed end moment as well as axial load. The B.S.C. literature gives a detailed account of the testing of joints subjected to axial load and moments in the chords.

The Mero joint consists of a cast steel ball into which the ends of the space grid members are screwed with a special end connector. The members are usually circular hollow sections. The Mero balls are made in a range of sizes to accommodate different member sizes. Up to 18 members can be connected at one joint. There is no eccentricity at the joints.

Other proprietary joint systems are available. Joints may also be made by site welding the individual members together on to cast steel spheres. This makes the construction expensive.

9.2.3 Analysis and design

Space grids are highly redundant and the analysis is carried out using a space frame program. Standard packages such as the I.C.L. space frame program are available. The frame may be analysed for pinned or fixed joints. Space frames constructed with proprietary joint systems should be treated as pin jointed. Preliminary sizes have to be assumed for the members in order to carry out the analysis. The members may then be designed and checked against the sizes originally assumed. The frame may be reanalysed and redesigned if necessary.

Some of the space frame programs available are limited in size. In these cases where the loading and arrangement of supports is symmetrical, only part of the frame need be analysed. This method permits the analysis of a large frame to be carried out. For a square grid, one-quarter of the frame is considered and for a rectangular grid, one-half of the frame. Rotational and linear restraints have to be introduced at the ends of members cut by the section planes. If members are sectioned length-wise, one half of the values of the structural properties, i.e. area, moments of inertia and torsion constant, are used in the analysis. The loading must also be halved on members or joints lying on the section planes. This process is shown in the example in Section 9.3.

The British Steel Corporation, Tubes Division, have developed a computer program for the analysis and design of space grids using the Nodus joint. This program determines the axial forces in all the members and the moments due to the joint eccentricities of the web members and to applied loads from cladding and services and distributes these into the chord members. For the members chosen, the program calculates the combined stress criterion from the interaction formula from Clause 14a of BS449:

$$f_c/p_c + f_{bc}/p_{bc} \leqslant 1.$$

The value of the left-hand side of the expression is printed and gives a measure of the efficiency of the member selected.

The B.S.C. literature also gives an approximate manual method of obtaining the axial forces in the members. This permits a preliminary design to be made first after which the grid can be analysed and the member size checked using the computer program.

If the space grid is considered as pin jointed the design of the members is carried out as follows.

(a) The top chord members are designed for axial load and bending due to the roof loads applied to the chords either through purlins or distributed along the member. The effective length must be estimated. This depends on the roof used for buckling in the plane of the top surface. For buckling perpendicular to the roof surface the effective length may be taken from 0.85–0.95 of the distance between centres of joints.

(b) The lower chord members are in tension where the weight of roof and space deck is greater than the uplift due to wind. This is generally the case. Local bending, however, will occur if loads from ceiling or services are supported on these members.

(c) The web members are in tension or compression and are designed accordingly. The effective length for compression members can be taken as 0.85 of the length between centres of nodes.

In general, all designs for joint system space grids make use of structural hollow sections for all members.

9.3 Preliminary design for a space grid

9.3.1 Specification
Make a preliminary design for a space grid for a roof 20 m square. The roof construction consists of steel deck, insulation board and three layers of felt. The total dead load including an allowance of 0.2 kN/m^2 for the space grid is 0.9 kN/m^2. The imposed load is 0.75 kN/m^2. The steelwork is exposed on the underside. The steel roof deck applies one way loading to the top chord members. The bottom chord members are to be designed to carry a concentrated load of 0.9 kN. This is a maintenance load and is not considered in the grid analysis. The design is to be in Grade 50 steel.

9.3.2 Arrangement of the space deck
The arrangement for the space deck is shown in Fig. 9.6(a). The grid is square on square offset with mansard edge. The grid is supported on columns around the edges located at the corners and every second bottom chord node point. The top and bottom subdivision by the chords is into squares with a side length of 2.5 m. The side walls are suitably braced to resist wind loading. The does not form part of the problem.

9.3.3 Approximate analysis and design
This is carried out to obtain the maximum forces in the space grid members

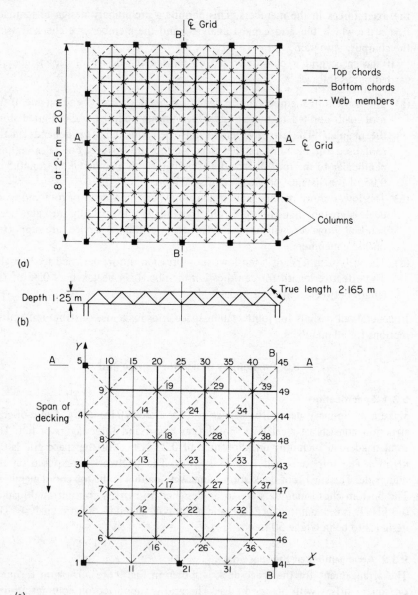

Figure 9.6 (a) Plan of grid. (b) Section. (c) One quarter of grid – used in the analysis.

from which the sizes of members to be used in the analysis can be determined. See [1].

Space frame grids, Part 1, Design. For a grid square on square offset supported regularly around all four edges.

T = total load on the grid = 1.65×20^2 = 660 kN

n = top chord module dimension   = 2.5 m

D = depth of grid   = 1.25 m

b = width of grid   = 20 m

$F_1 = F_2$ = chord force factors   = 0.08

Maximum chord force = $660 \times 2.5 \times 0.08/1.25$ = 105.6 kN

The design of the grid members follows:

(i) *bottom chord*

    Area = $105.6 \times 10/215$ = 4.9 cm^2

    Use 60.3 x 3.2 CHS: A = 5.74 cm^2.

(ii) *top chord*   In addition to the axial load, this is loaded by the one-way spanning deck.

    Chord vertical load = $2.5 \times 2.5 \times 1.65$ = 10.31 kN

    Moment   = $10.31 \times 2.5/8$ = 3.22 kN m

    Try 90 x 90 x 3.6 RHS where

      A = 12.4 cm^2, r = 3.52 cm, Z = 34.1 cm^3

    l/r_X = $0.9 \times 2500/34.1$ = 66

      p_c = 158 N/mm^2

    p_{bc} = 230 N/mm^2

      f_c = $105.6 \times 10/12.4$ = 85.2 N/mm^2

    f_{bc} = $3.22 \times 10^3/34.1$ = 94.4 N/mm^2

    $\dfrac{f_c}{p_c} + \dfrac{f_{bc}}{p_{bc}} = \dfrac{85.2}{158} + \dfrac{94.4}{230} = 0.54 + 0.41$ = 0.95

    Use 90 x 90 x 3.6 RHS.

(iii) *web members*

    Maximum reaction = $660 \times 2.5 \times 0.08 \times 8/20$ = 52.8 kN

    Length of web member = $(3 \times 1.25^2)^{0.5}$ = 2.17 m

    Maximum load in a web member

       = $52.8 \times 2.17/(1.25 \times 2)$ = 45.7 kN

Try 60.3 x 3.2 CHS. $A = 5.74 \text{ cm}^2,$ $r = 2.02 \text{ cm}$

$l/r = 0.9 \times 2170/20.2$ $= 96.7$

$p_c = 98 \text{ N/mm}^2$

$f_c = 45.7 \times 10/5.74$ $= 79.6 \text{ N/mm}^2$

Use 60.3 x 3.2 CHS.

The preliminary member sizes and the properties from the Constrado *Structural Steelwork Handbook* for structural hollow sections to be used in the analysis are given in Table 9.1.

Table 9.1   Member properties of analysis

Member	Section	Area	Moments of inertia		Torsional constant
		(cm^2)	$I_v(\text{cm}^4)$	$I_w(\text{cm}^4)$	$J\,(\text{cm}^4)$
Top chord	90 x 90 x 3.6 RHS	12.4	154	154	237
Bottom chord	60.3 x 3.2 CHS	5.74	23.5	23.5	46.9
Web members	60.3 x 3.2 CHS	5.74	23.5	23.5	46.9

9.3.4 Computer analysis

This is carried out using the I.C.L. space frame program [6]. The analysis is made for a pin-jointed frame with no eccentricity at the joints. Owing to symmetry of the grid and the loading only one-quarter of the frame need be considered in the analysis. The data for this space frame program are discussed and set out below.

(a) Joints and co-ordinates

The quarter frame used in the analysis and the joint numbering are shown in Fig. 9.6(c). The joint co-ordinates can be readily taken off this figure. The number of parameters for the quarter of the frame considered and the program limitations are:

Number of joints   = 49   Program limit = 100
Number of members   = 136   Program limit = 150
Maximum joint
  numbering difference = 10   Program limit = 15.

(b) Member properties

These are given in Table 9.1. Note that for the bottom chord members lying on the section planes AA and BB only one-half of the values for the complete member are used. Because the frame is pin jointed only the area is in fact used in the analysis.

(c) Restraints

The restraints at the supports and where members are cut by the section planes are listed in Table 9.2 using the I.C.L. space frame notation. Here:

L is a linear restraint + direction axis, e.g. *LX* is a linear restraint in the direction of the *X* axis

R is a rotational restraint + axis about which the restraint is applied e.g. *RZ* is a rotational restraint about the *Z* axis.

Table 9.2   Restraints

Joint no.	Restraint
45	LX, LY, RX, RY, RZ
15, 25, 35, 10, 20, 30, 40	LY, RX
42, 43, 44, 46, 47, 48, 49	LX, RY
5	LY, LZ, RX
41	LX, LZ, RY

The number of restraints required is 15. The program limitation is 80.

(d) Loading

The loads are applied by the one-way spanning roof deck across the top chord members and the sloping edge web members. The loads are as follows:

(i) Top chord joints 17, 18, 19, 27, 28, 29, 37, 38, 39

Load = 2.5 x 2.5 x 1.65   = 10.31 kN

This is a distributed load causing bending in the top chord members. The load will be applied as a point load at the nodes. The bending moment is calculated separately.

(ii) Top chord joints 7, 8, 9, 16, 26, 36

Sloping sides, length = $(2 \times 1.25^2)^{0.5}$ = 1.77 m

Dead load on slope = 0.9 x 1.77 x 2.5   = 3.98 kN

Imposed load on plan = 0.75 x 1.25 x 2.5   = 2.34 kN

Joint load = (10.31 + 3.98 + 2.34)/2   = 8.33 kN

(iii) Edge joints 2, 3, 4, 11, 21, 31

Joint load = (3.98 − 2.34)/2   = 3.16 kN

(iv) Edge joints 5, 41 Joint load   = 1.58 kN

(v) Edge joint 1

Joint load $(0.9 \times 0.884^2) + (0.75 \times 0.625^2)$ = 0.996 kN

(vi) Top chord joint 6

Joint load = $(1.65 \times 1.25^2) + (0.9 \times 0.884 \times 3.125)$
      $+ (0.75 \times 0.625 \times 3.125)$ = 6.53 kN

9.3.5 Computer results
The computer output is shown for the top chord and bottom chord members in Fig. 9.7(b) and (c), respectively. The maximum force in the web member in each unit only is given in Fig. 9.7(c).

9.3.6 Preliminary design of space grid members

(a) Top chord — 39–49, 39–40, 29–39, 38–39
       28–38, 38–48, 28–29, 29–30
       19–29, 18–28, 27–28, 37–38

Maximum force   = 102.3 kN compression

Moment   = 3.22 kN m

Adopt the section from the preliminary analysis.

Use 90 x 90 x 3.6 RHS.

Alternative section 114.3 x 3.6 CHS.

(b) Top chord — remaining members

Maximum force   = 57.6 kN compression

Moment   = 3.22 kN m

Try 80 x 80 x 3.6 RHS, where

$A = 10.9 \text{ cm}^2$, $r = 3.11 \text{ cm}$, $Z = 26.5 \text{ cm}^3$

$l/r = 0.9 \times 2500/31.1$ = 72.2

$p_c = 145 \text{ N/mm}^2$

$p_{bc} = 230 \text{ N/mm}^2$

$f_c = 57.6 \times 10/10.9$ $= 52.8 \text{ N/mm}^2$

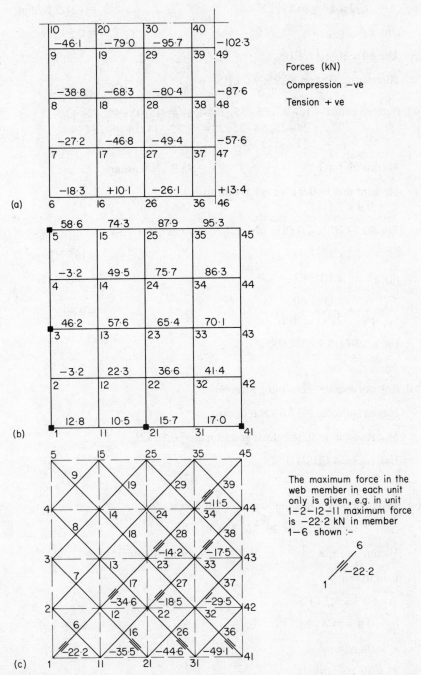

Forces (kN)
Compression −ve
Tension + ve

The maximum force in the web member in each unit only is given, e.g. in unit 1−2−12−11 maximum force is −22·2 kN in member 1−6 shown :-

Figure 9.7 (a) Top chord members. (b) Bottom chord members. (c) Web members.

$f_{bc} = 3.22 \times 10^3/26.5$ $= 121.5 \text{ N/mm}^2$

$f_c/p_c + f_{bc}/p_{bc}$ $= 0.89$

Use 80 x 80 x 3.6 RHS.

Alternative section 88.9 x 4.0 CHS.

(c) Bottom chord — 15–25, 25–35, 35–45, 44–45, 43–44, 42–43
      24–25, 34–35, 23–24, 33–34, 24–34, 34–44
      23–33, 33–43

Maximum force $= 95.3$ kN tension

Moment due to 0.9 kN load at mid span
  $= 0.9 \times 2.5/4$ $= 0.56$ kN m.

Try 60.3 CHS x 4.0 CHS where $A = 7.07 \text{ cm}^2$, $Z = 9.34 \text{ cm}^3$

$f_t = 95.3 \times 10/7.07$ $= 135 \text{ N/mm}^2$

$f_{bt} = 0.56 \times 10^3/9.34$ $= 60 \text{ N/mm}^2$

$\dfrac{f_t}{p_t} + \dfrac{f_{bt}}{p_{bt}} = \dfrac{135}{215} + \dfrac{60}{230}$ $= 0.89.$

Use 60.3 CHS x 4.0 CHS.

(d) Bottom chord — remaining members

Maximum force = 57.6 kN tension

Moment due to 0.9 kN load at mid span = 0.56 kN m

Use 48.3 x 4 CHS.

(e) Web members — meeting top chord nodes

6, 16, 26, 36, 7, 8, 9, 17

Maximum force $= 49.1$ kN compression

Length $= 2170$ mm

Try 60.3 x 3.2 CHS, $A = 5.74 \text{ cm}^2$, $r = 2.02$ cm

$l/r = 0.9 \times 2170/20.2$ $= 96.7$

$p_c = 98 \text{ N/mm}^2$

$f_c = 49.1 \times 10/5.74$ $= 85.5 \text{ N/mm}^2$

Use 60.3 x 3.2 CHS.

(f) Web members — remaining members

Maximum force = 29.5 kN compression

Try 48.3 x 3.2 CHS, $A = 4.53 \text{ cm}^2$, $r = 1.6$ cm

$l/r = 0.9 \times 2170/16$ = 122

$p_c = 65 \text{ N/mm}^2$

$f_c = 29.5 \times 10/4.53$ = 65.0 N/mm²

Use 48.3 x 3.2 CHS.

The proposed sections for the various members are shown in Fig. 9.8. A final design can only be carried out in conjunction with the grid joint manufacturer. If the Nodus joint is to be used, the frame should be analysed and designed using the British Steel Corporation program. The joint to be used will influence the changes, if any, that should be made in the sizes of members in the top and bottom chords and web.

9.4 Framed domes

9.4.1 Types and construction

The dome is formed by either curved members framing a surface of revolution or by straight members meeting at joints which lie on the surface. The most usual domical surface is part of a sphere covering an area circular in plan. However, a domed roof can be framed over other plan shapes.

Domes are classified according to the way in which the surface is framed. Many framing patterns are used but the main types are as follows:

(a) Schwedler or braced ribbed dome. This consists of ribs or meridional members converging at the top and parallel rings lying horizontal and spaced at equal arc lengths apart. The panels are braced by diagonal members. The dome may also be made with rigid joints without bracing. The top members usually meet in a compression ring. The bottom ring is in tension. A braced dome is shown in Fig. 9.9(a).

(b) Network and lamella domes. The main feature is parallel rings of members which are spaced at equal arc lengths apart. The rings are then interconnected by various bracing patterns. See Fig. 9.9(b). The Houston dome is a parallel lamella type of 200 m span.

(c) Grid dome. This is formed by a two- or three-way intersecting grid of arcs. The arcs are usually great circles. See Fig. 9.9(c).

(d) Geodesic dome. This system of dome construction was developed and patented by Buckminster Fuller. It is based on the icosahedron which is a solid with 20 faces, each an equilateral triangle. This solid is enclosed by a sphere touching the apicies of the triangles. The dome is formed from part of the sphere. Each equilateral triangle must be subdivided further to make

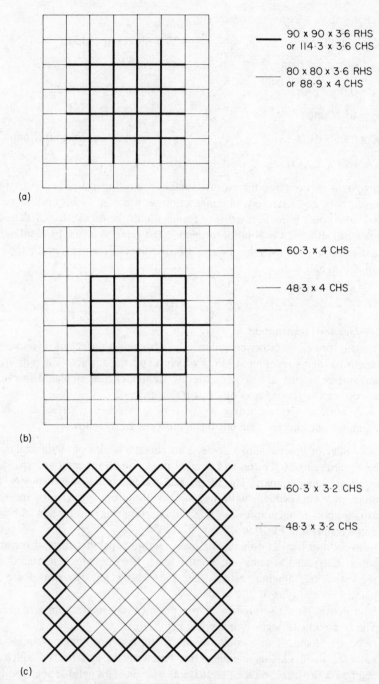

Figure 9.8   (a) Top chord members. (b) Bottom chord members. (c) Web members.

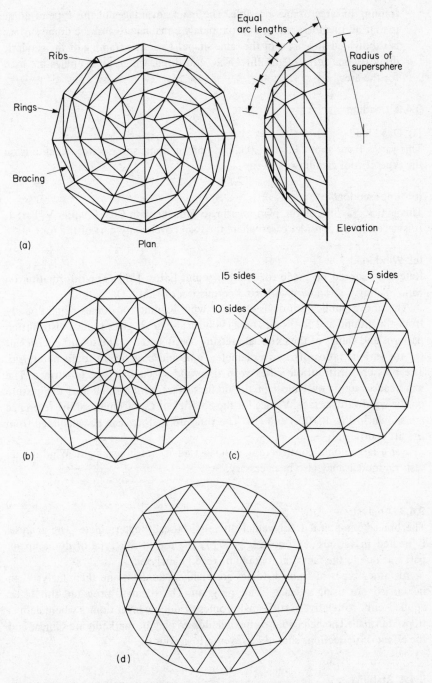

Figure 9.9 (a) Braced Schwedler dome. (b) Lamellar dome. (c) Network dome. (d) Three-way grid based on great circles.

framing of large domes possible. The main advantages of this type of dome is that all members are of approximately equal length and the dome surface is subdivided into areas of the same shape. The basic framing of the geodesic dome is one form of the three-way grid dome where great circles are used for the arcs.

9.4.2 Loading

(a) Dead load
This varies from about 0.5 to 1.0 kN/m^2 of the dome surface area depending on the type of roof cladding used.

(b) Imposed load
This is 0.75 kN/m3 of the plan area in accordance with CP3, Chapter V, Part 1. It is necessary to consider cases where the load covers only part of the roof.

(c) Wind load
Reference should be made to: Newberry and Eaton [11]. The information on wind loading is given here with kind permission of H.M.S.O.

The above publication deals with the wind loading for domes rising directly from the ground and for domes on a cylindrical base. It gives typical distribution patterns and values of the external pressure coefficient C_{pe} and the lift coefficient C_L for various ratios of rise to diameter, y/d, and base height to diameter, h/d. The pressure distribution varies with the wind speed and type of surface. The wind causes uplift over most of the dome surface with a small area of pressure on the windward surface. Values for the coefficient C_{pe} from the above reference are shown in Fig. 9.10(a) and (b). The total uplift force can be calculated from the lift coefficient.

For a large dome construction, specialist advice must be sought. Wind tunnel tests on models may also be necessary.

9.4.3 Analysis
The braced pin-pointed Schwedler dome is statically determinate. The analysis is treated in various textbooks (e.g. [7]). The use of this type of dome in the past was due to the fact that it could be readily analysed.

All other types of framed domes are highly redundant and their analysis can be carried out using a space frame program. The standard programs should be applied only to relatively stiff small domes or double layer domes where deflections are small. The behaviour of flexible domes may be markedly non-linear and the effect of deflections should be taken into account.

9.4.4 Stability
Flexible domes present a stability problem. Three distinct types of buckling can

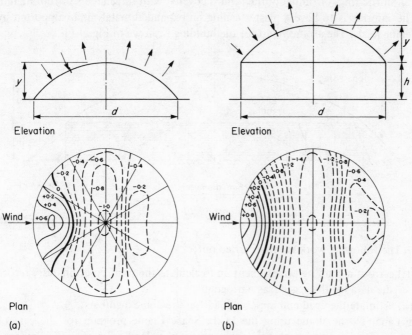

Figure 9.10   (a) Dome rising from the ground. Values of C_{pe} for y/d = ½, lift coefficient C_L = 0.5. (b) Dome on a cylindrical base. Values of C_{pe} for y/d = ½ and H/d = ½. Distributions and values of C_{pe} are reproduced with kind permission from [11].

occur in the framed dome. These are:

(a) general buckling where a large part of the surface buckles. Failures of this type have occurred where snow loads have only covered part of the surface;

(b) snap through or local buckling where one loaded node deflects and the dome curvature is reversed between adjacent nodes;

(c) member buckling where an individual member buckles as a strut under axial load. This is taken into account in the normal member design.

Stability of domes is discussed by Johnson [10].

9.5 Dome roof design

9.5.1 Specification

A circular exhibition building 25 m diameter is to be provided with a spherical framed dome roof. The roof eaves are to be 3 m above floor level and the rise of the dome 4 m. The side walls are to be vertical constructed in brickwork with the dome supports independent of the walls. The roof is timber deck carried on concentric rings of timber purlins and is covered with three layers of roofing felt. The interior is to have a plaster ceiling on expanded metal mesh supported by ceiling joists. The arrangement for the building is shown in Fig. 9.11.

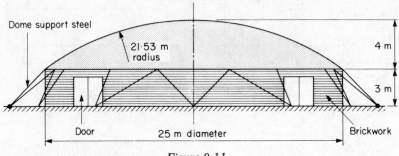

Dome support steel

21·53 m
radius

4 m

3 m

Door

25 m diameter

Brickwork

Figure 9.11

The following work is to be carried out:

(i) set out a framing arrangement and calculate the joint co-ordinates. Prepare the data for the space frame program;
(ii) estimate the dead and imposed loads on the dome members;
(iii) analyse the dome using the I.C.L. Space Frame program for the case of imposed loading over the whole roof. Other load cases of partial imposed loading and wind loading would have to be considered in a complete design;
(iv) make a preliminary design for the dome structure based on the analysis in (iii) using rectangular hollow sections for the members.

9.5.2 Framing arrangement and space frame program input data

(a) Framing arrangement and joint co-ordinates

The lamella dome arrangement shown in Fig. 9.12 will be adopted. The radius of the supersphere is 21.53 m. The top compression ring is 1.6 m diameter and the intermediate ring is midway between the apex and the tension ring at the eaves. The maximum span for purlins is about 3 m. All dome members and purlins are straight.

Because of symmetry of the frame and of the particular loading case only one-quarter of the dome need be considered in the analysis. The quadrant AB is selected and the joint numbering, co-ordinates and member lengths are shown in Fig. 9.13. The co-ordinates are listed in Table 9.3(a). The quadrant has 21 joints and 34 members.

(b) Member connection data

This is given in Table 9.3(b). This also gives the section properties references to be used.

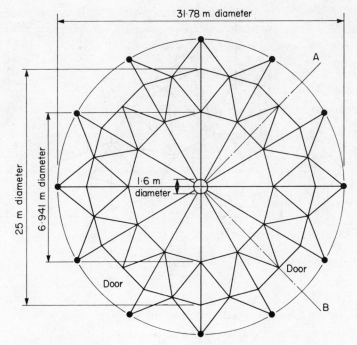

31·78 m diameter

A

25 m diameter

6·941 m diameter

1·6 m
diameter

Door

Door

B

Plan — Dome framing

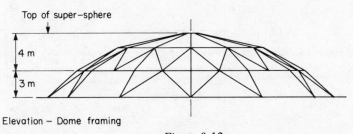

Top of super—sphere

4 m

3 m

Elevation — Dome framing

Figure 9.12

(c) Section properties

The sizes to be used in the analysis must first be selected from experience. These will be checked when the final design is made. The sizes to be used are given in Table 9.4. The properties reference to the individual members is given in Table 9.3(b).

(d) Restraints

The restraints at the joints on the section planes to give continuity are listed in Table 9.5. There are 32 restraints.

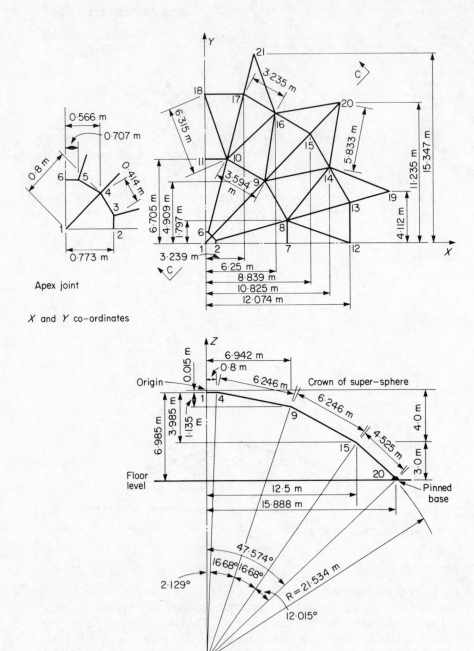

Apex joint

X and Y co-ordinates

Section C–C Z– co-ordinates

Figure 9.13

Table 9.3 Joint coordinates, member connection and section properties reference

(a) Joint co-ordinates

Joint	X	Y	Z	Joint	X	Y	Z
1	0.0	0.0	0.0	12	12.047	0.0	−3.985
2	0.773	0.0	0.0	13	12.047	3.235	−3.985
3	0.773	0.207	0.0	14	10.825	6.25	−3.985
4	0.566	0.566	0.0	15	8.839	8.839	−3.985
5	0.207	0.773	0.0	16	6.25	10.825	−3.985
6	0.0	0.773	0.0	17	3.235	12.047	−3.985
7	6.705	0.0	−1.135	18	0.0	12.047	−3.985
8	6.705	1.797	−1.135	19	15.347	4.112	−6.985
9	4.909	4.909	−1.135	20	11.235	11.235	−6.985
10	1.797	6.705	−1.135	21	4.112	15.347	−6.985
11	0.0	6.705	−1.135				

(b) Member connection and section properties reference

Joint	Joint	Reference	Joint	Joint	Reference	Joint	Joint	Reference
1	4	A	8	12	B	15	16	A
2	3	B	8	13	A	16	17	A
3	4	B	8	14	B	17	18	A
4	5	B	9	14	B	13	19	A
5	6	B	9	15	A	14	19	A
3	8	A	9	16	B	14	20	A
4	9	A	10	16	B	15	20	A
5	10	A	10	17	A	16	20	A
7	8	B	10	18	B	16	21	A
8	9	B	12	13	A	17	21	A
9	10	B	13	14	A			
10	11	B	14	15	A			

Table 9.4 Member properties for analysis

| Reference | Member | Area | Moments of inertia | | Torsional constant |
| | | | I_v | I_w | |
		(cm^2)	(cm^4)	(cm^4)	J (cm^4)
A	150 x 100 x 6.3 RHS	29.7	910	479	985
B	100 x 100 x 5 RHS	18.9	283	283	439

Table 9.5   Restraints

Joint	Linear			Rotational		
	X	Y	Z	X	Y	Z
1	LX	LY		RX	RY	RZ
2		LY		RX		RZ
6	LX				RY	RZ
7		LY		RX		RZ
11	LX				RY	RZ
12		LY		RX		RZ
18	LX				RY	RZ
19	LX	LY	LZ			
20	LX	LY	LZ			
21	LX	LY	LZ			

9.5.3 Loading

The arrangement of the purlins is shown in Fig. 9.14. The roof load and the loads on the members and joints are calculated below. These form the loading data for the computer program.

(a) *Roof loads per unit area*   (kN/m^2)

    Roof — Timber purlins 175 x 50 at 1041 mm centres   0.061

                Timber deck 25 mm thick   0.175

                Three layers of roofing felt   0.195

    Ceiling — Joists 75 x 50 at 500 mm centres   0.053

                Expanded metal and plaster   0.25

    Dome steelwork   0.2

    Total dead load — on slope area   0.934

    Imposed load — on plan area   0.75

(b) Distributed loads on members

    Members 3–8, 4–9, 5–10

    Imposed load on slope = 0.75 x 6.142/6.246   = 0.738 kN/m^2

    Total load   = 0.93 + 0.738   = 1.672 kN/m^2

    Ends 3, 4, 5   0.414 x 1.672   = 0.692 kN/m

    Ends 8, 9, 10   3.594 x 1.672   = 6.009 kN/m

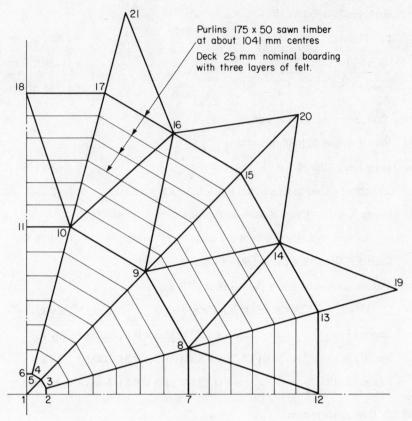

Figure 9.14

Members 8–13, 9–15, 10–17

Imposed load on slope = 0.75 x 5.558/6.246		= 0.667 kN/m²
Total load	= 0.934 + 0.667	= 1.601 kN/m²
Ends, 9, 10		= 0
Ends, 13, 15, 17		= 5.755 kN/m

Members 8–14, 9–14, 9–16, 10–16

Total load		= 1.601 kN/m
Ends 8, 9, 10	3.594 x 1.601/2	= 2.877 kN/m
Ends 14, 16	3.235 x 1.601/2	= 2.59 kN/m

Members 8–12, 10–18

Total load	= 1.601 kN/m
Ends 8, 10	= 1.439 kN/m
Ends 12, 18	= 1.245 kN/m

(c) Concentrated loads at joints

Joint 1 carries ½ load of top ring

Total load = 0.934 + 0.75	= 1.684 kN/m^2
On part of dome analysed = ¼ x ½ x 0.8^2 x π x 1.684	= 0.42 kN

Joints 3, 4, 5 – Carry ½ load on top ring

$$= ½ \times ½ \times 0.8^2 \times \pi \times 1.684 \qquad = 0.14 \text{ kN}$$

Joint 15 Allowance for weight of dome steelwork

Assume weight of dome steelwork is 23.3 kg/m

(3.235 + ½ x 4.525 + 5.833)23.3 x 9.81/1000	= 2.59 kN
Joints 14, 16	(5.833 + 3.235)23.3 x 9.81/1000 = 2.07 kN
Joints 13, 17	½[4.525 + 5.833] 23.3 x 9.81/1000 = 1.18 kN
Joints 12, 18	½ x 3.235 x 23.3 x 9.81/1000 = 0.37 kN

9.5.4 Computer results
The results of the computer analysis giving actions in the dome members are given in Table 9.6. One member from each group is selected. The direct shears have not been listed.

9.5.5 Preliminary design

(a) Support ribs Member 13–19   End 13

Axial compression	= 80.33 kN
Moments M_v	= 8.06 kN m
$\qquad M_w$	= 0.18 kN m
Torsion	= 0.29 kN m

Trial section 200 x 100 x 5 RHS

A = 28.9 cm^2	Z_{YY} = 102 cm^2	r_{XX} = 7.23 cm
Z_{XX} = 151 cm^3	r_{YY} = 4.2 cm	C = 172 cm^3

Table 9.6   Member actions from computer analysis

Member		Axial load (kN)	M_V (kN m)	M_W (kN m)	Torsion (kN m)
Top	1	+26.9	− 6.27	−	−
ribs	4	−26.9	6.61	−	−
Top	4	+43.1	− 3.35	−1.07	−1.7
ring	5	−43.1	3.05	−0.59	+1.7
Ribs	4	+41.4	− 5.06	−	−
	9	−45.2	17.68	−	−
Ring	10	+106.03	− 0.69	−	−
	11	−106.03	+ 0.69	−	−
Ribs	10	71.41	− 5.14	0.36	0.47
	17	−79.59	5.46	0.31	−0.47
Diagonals	9	18.08	− 9.24	0.25	0.24
	16	−25.37	6.79	−0.18	−0.24
Ring	14	−15.61	1.93	−0.2	−0.89
	15	15.61	− 0.98	−0.35	+0.89
	17	− 7.88	− 3.08	−0.94	0.61
	18	7.88	− 7.41	−1.7	−0.61
Support	13	80.33	− 8.06	0.18	−0.29
rib	19	−80.33	0.0	0.0	+0.29
Support	14	36.4	− 5.89	−2.69	−1.27
diagonal	20	−36.4	0.0	0.0	1.27

Sign convention — Member axes

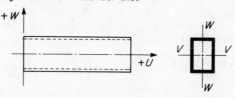

The member is not supported laterally

$l/r_{YY} = 0.85 \times 4530/42$ $= 91.7$

   $p_c = $ 89 N/mm^2

$p_{bc} = 165$ N/mm^2

Axial stress $f_c = 80.3 \times 10/28.9$ $= 27.8$ N/mm^2

Bending stress XX-axis $f_{bc} = 8.06 \times 10^3/151$ $= 53.4$ N/mm^2

         YY-axis $f_{bc} = 0.18 \times 10^3/102$ $=$ 1.8 N/mm^2

Torsion stress $f_s = 0.29 \times 10^3/172$ $= 1.7 \text{ N/mm}^2$

Combined $27.8/89 + 55.2/165$ $= 0.65$

Make support members 200 x 100 x 5 RHS

(b) Ribs   Member 4–9

The loading is shown on Fig. 9.15. The value of the maxium sagging moment is determined. The point of zero shear is given by:

$$5.55 = 0.69x + (6.01 - 0.69)(x/6.26)(x/2)$$

which reduces to give

$x^2 + 1.6x - 12.9$ $= 0$

from which x $= 2.88$ m.

$M_p = (5.55 \times 2.88) - 5.06 - 0.69 \times 2.88^2/2 - 2.45$

$\times 2.88^2/6$ $= 4.67$ kN m.

Design the member for the actions at end 9.

Trial section 200 x 100 x 5 RHS.

$l/r_Y = 0.85 \times 6250/42$ $= 127$

$p_c = 55 \text{ N/mm}^2$

$p_{bc} = 165 \text{ N/mm}^2$

$f_c = 45.2 \times 10/28.9$ $= 15.6 \text{ N/mm}^2$

$f_{bc} = 17.68 \times 10^3/151$ $= 117 \text{ N/mm}^2$.

Combined:

$(15.6/55) + (117.1/165)$ $= 0.99$.

Make all the ribs and supports 200 x 100 x 5 RHS.

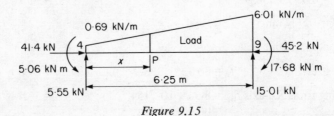

Figure 9.15

(c) Diagonals Member 9–16   End 9

Axial compression $= 18.08$ kN

Moments $M_v = 9.24$ kN m

$M_w = 0.25$ kN m

Torsion $= 0.24$ kN m

Trial section 150 x 100 x 5 RHS.

$A = 23.9$ cm^2, $Z_{YY} = 79.1$ cm^3, $r_{XX} = 5.59$ cm

$Z_{XX} = 99.5$ cm^3, $r_{YY} = 4.07$ cm, $C = 127$ cm^3

$l/r_{YY} = 0.75 \times 6315/40.7$, $= 116.4$

$p_c = 62$ N/mm^2

$f_c = 18.08 \times 10/23.9$ $= 7.6$ N/mm^2

XX axis $f_{bc} = 9.24 \times 10^3/99.5$ $= 92.9$ N/mm^2

YY axis $f_{bc} = 0.25 \times 10^3/79.1$ $= 3.2$ N/mm^2

Combined:

$(7.6/62) + (96.1/165)$ $= 0.704.$

Make all diagonals 150 x 100 x 5 RHS.

(d) Support diagonals, Member 14–20   End 14

Axial compression $= 36.4$ kN

Moments $M_v = 5.89$ kN m

$M_w = 2.69$ kN m

Torsion $= 1.27$ kN m

Trial section 150 x 100 x 5 RHS.

$l/r_{YY} = 0.85 \times 5833/40.7$ $= 121.8$

$p_c = 58$ N/mm^2

$p_{bc} = 165$ N/mm^2

$f_c = 36.4 \times 10/23.9$ $= 15.2$ N/mm^2

XX axis $f_{bc} = 5.89 \times 10^3/99.5$ $= 59.2$ N/mm^2

YY axis $f_{bc} = 2.69 \times 10^3/79.1$ $= 34.0$ N/mm^2

Combined:

$(15.2/58) + (93.2/165)$ $= 0.83.$

Torsion stress $= 1.27 \times 10^3/127$ $= 10 \text{ N/mm}^2$

Make all supports diagonals 150 x 100 x 5 RHS.

(e) Compression rings, Member 10–11

Axial compression $= 106.03 \text{ kN}$

Moment $M_V = 0.69 \text{ kN m}$

Trial section 100 x 100 x 5 RHS.

$A = 18.9 \text{ cm}^2,$ $Z_{XX} = 56.6 \text{ cm}^3,$ $r = 3.87 \text{ cm}$

$l/r = 0.85 \times 3594/38.7$ $= 78.9$

$p_c = 105 \text{ N/mm}^2$

$p_{bc} = 165 \text{ N/mm}^2$

$f_c = 106.03 \times 10/18.9$ $= 56.1 \text{ N/mm}^2$

$f_{bc} = 0.69 \times 10^3/56.6$ $= 12.2 \text{ N/mm}^2$

Combined:

$(56.1/105) + (12.2/165)$ $= 0.61. \text{ Safe.}$

For uniformity the top ring will be made 200 x 100 x 5 RHS and the intermediate ring will be made 150 x 100 x 5 RHS.

(f) Tension ring, Member 17–18   End 18

Axial tension $= 7.88 \text{ kN}$

Moments $M_V = 7.41 \text{ kN m}$

                 $M_W = 1.7 \text{ kN m}$

Try 150 x 100 x 5 RHS.

$f_t = 7.88 \times 10/23.9$ $= 3.3 \text{ N/mm}^2$

XX axis $f_{bt} = 7.41 \times 10^3/99.5$ $= 74.5 \text{ N/mm}^2$

YY axis $f_{bt} = 1.7 \times 10^3/79.1$ $= 21.5 \text{ N/mm}^2$

Combined:

$(3.3/155) + (96/165)$ $= 0.6.$

Make all members in ring 150 x 100 x 5 RHS.

(g) Summary of member sizes

The member sizes from the preliminary analysis and design are summarized in Table 9.6.

Table 9.6 Member sizes

Location	Members	Section
Supports	13–19, 15–20, 17–21	200 x 100 x 5 RHS
Support diagonals	14–19, 14–20, 16–20, 16–21	150 x 100 x 5 RHS
Main ribs	8–13, 9–15, 10–17 3–8, 4–9, 5–10, 1–4	200 x 100 x 5 RHS
Diagonals	8–12, 8–14, 9–14, 9–16 10–16, 10–18	150 x 100 x 5 RHS
Top ring	2–3, 3–4, 4–5, 5–6	200 x 100 x 5 RHS
Intermediate ring	7–8, 8–9, 9–10, 10–11	150 x 100 x 5 RHS
Lower ring	12–13, 13–14, 14–15, 15–16, 16–17, 17–18	150 x 100 x 5 RHS

(h) Construction

The dome can be divided into sections for shop fabrication. Field connections can be bolted flanged joints made with high-strength friction grip bolts.

9.6 Coursework exercises

1. Make a preliminary design for a space grid for a roof 24 m square. Use grid construction square on larger square set diagonally with cornice edge. Column supports are to be located at 6 m centres on all sides. The total dead load is 0.7 kN/m^2 and the imposed load is 0.75 kN/m^2. The work to be carried out is:

   (a) set out the arrangement of the space grid;
   (b) make an approximate analysis and design using the method set out in the British Steel Corporation publication on space grids;
   (c) analyse the grid using a space frame program for pinned joints;
   (d) make a preliminary design for the grid members.

2. A foyer in a building is to be provided with a domical roof. The dome diameter is 12 m and the rise is 3 m. It is proposed to frame the dome by a three-way grid using great circles. The dome covering will be made by special prefabricated roof units. The dome is to have a tension ring at the springing and it is carried on a reinforced concrete structure.

   (a) Set out the dome in plan and elevation and calculate the joint co-ordinates.

(b) Calculate the loading on the members. The dead load is 0.6 kN/m^2 and the imposed load is 0.75 kN/m^2.

(c) Analyse the dome using a space frame program.

(d) Make a preliminary design for the dome members.

References and further reading

[1] British Steel Corporation Tubes Division (1974). Nodus – Space Frame Grids – Part 1 Design, Part 2 Analysis, Part 3 Construction. Croydon.

[2] British Steel Corporation (1972). *Building with Steel*. Theme Space Structures. London.

[3] Makowski, Z. S. (1965). *Steel Space Structures*. Michael Joseph, London.

[4] Davis, R. M. (Ed.) (1966). *Space Structures*. Blackwell Scientific, Oxford Publications.

[5] *Steel Designers Manual* (1972). Crosby Lockwood, London.

[6] I.C.L. (1967). *Analysis of Space Frames*. International Computers Limited.

[7] Benjamin, B. S. (1963). *The analysis of braced domes*. Asia Publishing House, London.

[8] Marks, R. W. (1960). *The dynaxion world of Buckminster Fuller*. Reinhold, New York.

[9] Pugh, A. (1976). *Polyhedra – A visual approach*. University of California Press.

[10] Johnson, B. G. (Ed.) (1976). *Guide to stability design criteria for metal structures*, 3rd edn. New York, John Wiley.

[11] Newberry, C. W. and Eaton, K. J. (1974). *Wind Loading Handbook*. Building Research Establishment, H.M.S.O., London.

10 Multi-storey buildings

10.1 General considerations

It is instructive to classify multi-storey buildings and discuss the important structural principles involved in their design and construction. The special problems discussed will be mainly those concerned with the very tall buildings of which a selection of types are considered. In the United States structures over 400 m height with more than 100 storeys have been built. Fig. 10.1(a) shows a medium rise building and (b) a tall building.

Tall buildings are mainly constructed in city centres where land is in short supply and high population density coupled with high land prices and rents make their use economical. In provision of housing, one tall building can replace a large area of low-rise buildings which can then be developed as open parkland. Tall buildings are used for offices, banks, hotels, flats, schools, hospitals, department stores, etc.

Architects and engineers planning a tall building need to consider the following general constraints on the design.

(a) Building regulations and planning laws for the city concerned. Sometimes the maximum building height is limited.
(b) Intended occupancy. This governs the floor loading and influences the arrangement adopted. For example, in flats the division of floor space can be the same on each floor and vertical load-bearing walls throughout the height of the building can be introduced. Whereas in office buildings the division of space between core and perimeter is made with moveable partitions to give maximum flexibility. See Fig. 10.1(b).
(c) The transport of people is primarily vertical and the provision of a core containing stairs and lifts has a great influence on design. The design of tall buildings only became possible following the development of the electric hoist.
(d) Fire protection of the structural frame is mandatory as is provision of separate fireproof compartments for the lifts and stairs. The building must be able to be evacuated as rapidly as possible. The design must comply with relevant regulations.
(e) Heating is a necessity and air conditioning is often provided. This requires space between floor slabs and suspended ceilings and in the curtain walls to accommodate ducts and pipes.

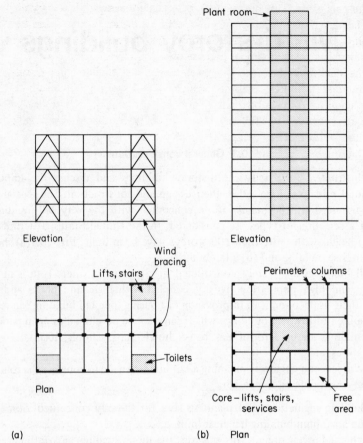

Figure 10.1   (a) Medium-rise building. (b) Tall building.

(f) The provision of services is an important part of the design that must be considered from the planning stage. Services can be included in prefabricated wall and floor units during manufacture.

10.2 Structural design considerations

In the structural engineering sense the multi-storey building may be defined as tall when the horizontal loading due to wind or seismic loading becomes an important consideration in the design. This is particularly so with modern buildings clad with light curtain walling and using lightweight fire protection and internal partitioning. In these structures there is no great mass of dead loads to provide stability.

The frame must be stiff enough to restrict the deflection to an acceptable value. This is usually limited to 1/500 of height to prevent sway causing anxiety to the occupants. This limitation is of prime importance with very tall buildings and has led to the development of special structural forms as the tube type of building described below. Multi-storey rigid frame construction alone is not suitable for very tall buildings because of excessive deflection.

The steel frame building with steel deck floors, prefabricated cladding panels, demountable internal partitions, suspended ceilings and lightweight fire protection lends itself to industrialized building techniques. The advantages of this type of construction are accuracy of shop fabrication of units and speed of erection which requires minimum site labour with few specialized skills.

The foundations can be expensive depending on soil conditions because heavy loads are delivered onto a small area. Individual footings with grillages to spread loads are used in some cases. Often in poor conditions cellular rafts or multi-storey basement foundations are provided. These may bear on to the soil or be supported on cylinder piles. Basements also serve as car parks.

Very often the erection of the structure has to be carried out on a restricted site. This influences the design and limits the size of components to be prefabricated. Buildings in the low- to medium-rise range can be erected with independent tower cranes. However, for tall buildings, erection must be carried out using the building itself. Here core-type structures are very suitable as the concrete core can be constructed first to full height by slipforming and this can be used to erect the steelwork. This is shown below in Figs. 10.4 and 10.5.

It is difficult to compare the costs of the different structural systems in use. Comparisons on the basis of frame cost only can be very misleading. All the relevant factors affecting capital cost and operating cost should be considered. These costs include the frame, foundations, flooring, cladding, types of partitioning, fire protection, air conditioning, other services, operating costs and maintenance. Particular contractors may specialize in certain types of construction and so can offer very competitive prices for that type. It is probably virtually impossible to say in advance which system will be the most economical in a given situation. All the various systems continue to be used with no one type predominating.

10.3 Structural systems

The main structural systems used are discussed below. The classification into the various types is based primarily on the structural form or method adopted to resist horizontal loading. The second classification is concerned with the method of construction used. Buildings may consist of a combination of various types noted.

This section is concerned with steel framed buildings. However, structural steel acting in conjunction with reinforced concrete elements occurs in some of the structures discussed where concrete shear walls and cores are used to provide stability. Design of the reinforced concrete elements is not discussed here.

10.3.1 Braced structures

The bracing forming a vertical cantilever resists the horizontal loading. The simple method of design involving only manual analysis may be used for the whole structure for buildings braced with one cantilever truss as shown in Fig. 10.2(a). A tall structure braced in one narrow bay may deflect too much. The deflection can be reduced by providing horizontal bracing at the top and at intermediate floors if necessary. This brings the outside columns into action in resisting horizontal load. The frame is now statically indeterminate but can be analysed using a computer program. See Fig. 10.2(b).

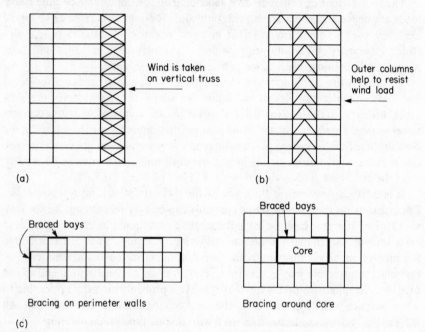

Figure 10.2 (a) Vertical bracing. (b) Vertical and horizontal bracing. (c) Location of bracing in plan.

The location of the braced bays can present problems as the bracing interferes with glazing and door openings. The bracing is usually located on perimeter walls, e.g. fully clad end walls or on the sides of a central core enclosing lift shaft and stairwell. This is shown in Fig. 10.2(c).

Construction is simple with site-bolted joints using black bolts. The floors act as horizontal beams to carry wind load to the vertical bracing trusses. The floors may require horizontal bracing in some cases depending on construction.

10.3.2 Shear wall structures

The shear wall structure consists of an arrangement of rigid vertical walls in two directions at right angles with the walls carrying vertical and horizontal loads.

This type of construction is more applicable to reinforced concrete and load-bearing brickwork buildings. The braced bay in the steel structure is equivalent to the shear wall. The braced bays in Fig. 10.2(c) could be replaced by concrete shear walls.

Concrete shear walls are often used to brace steel structures. The shear walls may be located at the ends or sides or around the core. The concrete core walls also form a fireproof compartment for stairs and lifts. The core may also be external to the building. If the arrangement of shear walls or the core is placed asymmetrically in the building plan they will be subjected to torsion as well as shear due to horizontal loading. The shear walls provide stability for the structure. They must not be removed or substantially perforated in future alterations. Some shear wall and core arrangements are shown in Fig. 10.3.

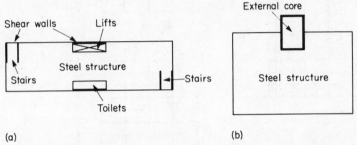

(a)   (b)

Figure 10.3   (a) Plan — steel framing with reinforced concrete shear walls. (b) Plan — steel framing with asymmetrical core.

10.3.3 Core structures

Many of the important features for this type of construction have already been mentioned. The main features are briefly summarized below

(a) The structural arrangement gives the following advantages:

    (i) the floor space between core and perimeter columns is free giving maximum flexibility using lightweight partitions;

    (ii) the lifts and staircases require a rigidly constructed fire-resistant shaft. The core provides this requirement;

    (iii) there is no bracing in the perimeter bays so the facade treatment can be uniform over all faces.

(b) The structural action is clearly expressed in that the core is designed to resist the wind loading which is transmitted to it by the floors. The floor plan may be square, rectangular, triangular, circular, etc. The floor steel may be supported in three ways. These are

    (i) supported on the core and perimeter columns, see Figs. 10.1(b) and 10.4(a);

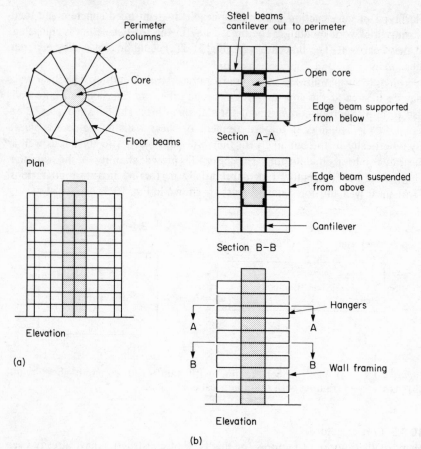

Figure 10.4 (a) Circular core structure with perimeter columns. (b) Floors cantilevered from core.

(ii) cantilevered out from the core, see below;

(iii) suspended from a girder at the top of the core. This is discussed in Section 10.3.4. The core should be located at the geometric centre of the plan.

(c) Construction is rapid using slip forming for the core. The core is then used to erect the building. The erection can be carried out within the area of the building.

The core may be of open or closed construction. The closed cores of box or tubular form are designed as vertical cantilevers. Open cores of angle, channel or H section are designed as connected cantilever shear walls. Cores may be constructed in reinforced concrete or composite steel and concrete construction with steel columns at the corners or ends. The design of cores is not considered further here.

Where floor beams are cantilevered out from the core, these are usually built as a storey-high frame with the beams on adjacent floors spanning at right angles to each other. This avoids the need for beams to interpenetrate each other at right angles. Edge and floor beams on one floor are then suspended or supported on struts from the cantilever beams on the other floor. See Fig. 10.4(b).

10.3.4 Suspended structures

In this case an umbrella girder is provided at the top of the core from which supporting members for the outer ends of the floor beams are suspended. In tall structures additional umbrella girders may be introduced at intermediate points in the height of the structure. A suspended structure is shown in Fig. 10.5.

All loads both vertical and horizontal are carried on the core in suspended structures. Sections, flats and round bars in high-strength steel or cables are used for the suspension members. The size of the perimeter hangers is kept to a minimum. The time for erection can be shorter than with a conventional structure built upwards. However, this form of construction requires special techniques and should only be carried out by experienced contractors.

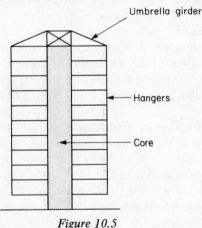

Figure 10.5

10.3.5 Rigid frame structures

In this type of structure, the horizontal load is resisted by bending in the beams and columns. The columns, particularly in the lower storeys, must resist heavy moments so will require a larger section than in structures where the wind loading is taken by bracing or on a core.

The frame is generally only rigid in one direction and is braced in the other direction. Such a structure is shown in Fig. 10.6. Connections between beams and columns must be capable of resisting moment. These must be site welded or made using high-strength friction grip bolts. Site-welded joints are only possible with careful fabrication and erection. This adds to the cost of the structure.

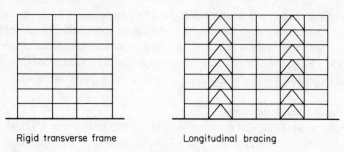

Rigid transverse frame   Longitudinal bracing

Figure 10.6

The bolted joints with HSFG bolts generally require haunched ends on the beams to obtain sufficient lever arm to make the moment-resisting joint.

The rigid frame structure deflects more than the braced structures. The deflection is made up of sway deflection in each storey plus a cantilever deflection of the whole frame. The deflection is given as part of the output of the computer analysis. Because of the excessive deflection, rigid frame structures are generally only used for low- and medium-rise buildings.

10.3.6 Tube structures

This type of structure was developed by Dr Fazlur Khan of the USA for very tall buildings over 30–40 storeys in height. If the core type of structure is used for these buildings, the core dimensions are too small compared with the height to limit the deflection satisfactorily at the top. The deflection is usually limited to 1/500 of height.

The perimeter walls are so constructed that the whole structure forms a rigid tube which is used as the means to resist horizontal load. The framed tube as shown in Fig. 10.7(a) consists of closely spaced exterior columns 1 to 3 m apart tied at each floor level with deep floor beams. This creates a hollow tube with small perforations for windows. This eliminates the curtain wall and mullion facade used on multi-storey buildings. The braced tube shown in Fig. 10.7(c) is also used.

In the single-tube structure the perimeter walls carry all the horizontal load and part of the vertical load. Internal columns and/or an internal core if provided carry vertical loads only. The tube in tube system shown in Fig. 10.7(d) is also used. Here the core is designed to carry part of the horizontal load. Very tall, stiff structures have been designed on the bundled tube system shown in Fig. 10.7(e) which consists of a number of tubes constructed together. This reduces the shear lag problem which would be serious if a single tube were used in this case.

The analysis of a tube structure may be carried out on a large space frame program. The main feature shown up in the analysis for horizontal load is the drop in the load taken by columns in the flange faces. This is caused by shear lag in the beam column frames as shown in Fig. 10.7(b). Simple beam theory

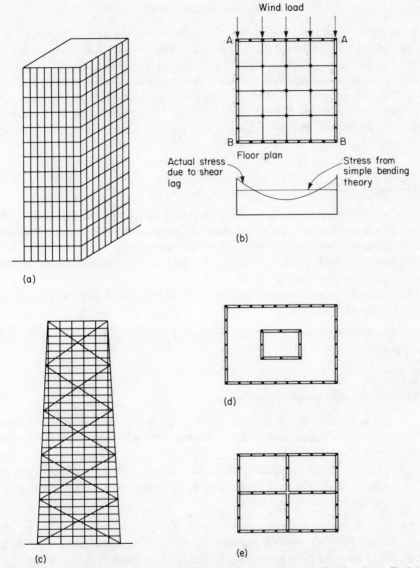

Figure 10.7 (a) Framed tube. (b) Floor plan and stress distribution in walls AA and BB. (c) Braced tube. (d) Plan − tube in tube arrangement. (e) Plan − bundled tube arrangement.

would give uniform loading in these columns. Dr Fazlur Khan has developed preliminary methods of analysis which take shear lag into account. The student should consult references given for further information [1, 4].

This type of structural form is very efficient and results in a considerable saving in material used when compared with conventional designs.

10.4 Constructional details

In order to idealize the structure for analyses and design and estimate the loading on the constituent plane frames it is necessary to have a knowledge of the various forms of construction used for the roof, floors and walls as well as the types of structural members used for the beams, columns and bracing. Brief details are set out below with sketches to show some common forms of construction used. Many proprietary building units and systems are available. The reader should consult manufacturers literature for further information. See also [3].

(a) Roofs

The main types of construction used for flat roofs are:

(i) cast *in situ* or precast concrete slabs carried on steel beams. The cast *in situ* slab may be designed for composite action with the steel beams. The finish is usually screed with asphalt topping and chippings;

(ii) steel deck with insulation board and three layers of roofing felt. The deck is carried on steel beams;

(iii) composite *in situ* concrete on steel deck. The steel deck serves as permanent formwork.

Suspended fire-resistant ceilings are required. Construction is shown in Fig. 10.8 (a) and (b).

(b) Floors

The main types of construction used are:

(i) cast *in situ* or precast concrete slabs supported on beams or open lattice girders. The design may provide for composite action between the cast *in situ* slab and the steel beams;

(ii) composite *in situ* concrete on steel deck floor supported on steel beams or lattice girders. The steel deck serves as permanent formwork. This type of floor lends itself to prefabrication.

Lattice or open web girders are convenient as they provide access for the insertion of services. The floor finish is generally screed and tiles. A fire-resistant ceiling is suspended from the floor beams or slab. Floor construction is shown in Fig. 10.8 (b).

(c) Walls

Walls in steel-framed buildings may be classified as follows:

(i) structural shear walls located in bays on the perimeter or around cores. These are of reinforced concrete or composite construction incorporating steel columns and are designed to resist horizontal wind loading as well as vertical loads;

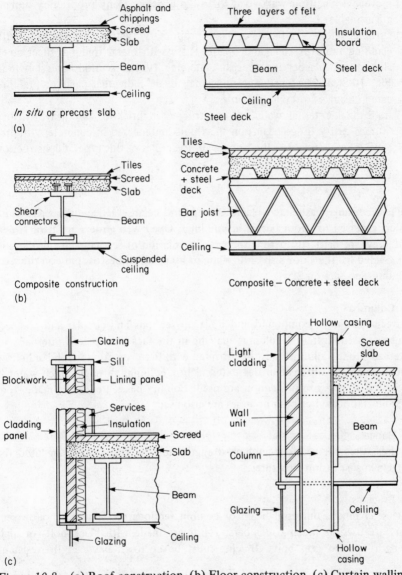

Figure 10.8   (a) Roof construction. (b) Floor construction. (c) Curtain walling.

(ii) non-load-bearing permanent division and fire-resistant walls. These are usually constructed in bricks or lightweight blocks. Fire-resistant walls are required around stairwells and liftshafts and to form the permanent fire compartment walls in buildings of large floor area. These walls are also needed to protect steel bracing members in braced bays;

(iii) moveable partition walls used for room division. Many proprietary designs are available;

(iv) curtain walling on facades. Many proprietary systems are available with cladding units in clear and opaque glass with metal framing, sheet steel, aluminium, plastic and precast concrete. Typical sections are shown in Fig. 10.8(c). The curtain walling may be in individual parts or prefabricated in single-storey height units;

(v) traditional external walls are constructed in brick of two leaves with a 50 mm cavity. Edge beams on the floors are encased in concrete with projecting lintels to carry the brick wall. Examples of this type of construction are shown in Fig. 4.24.

(d) Floor beams

Universal beams, castellated beams, compound beams, lattice girders and plate girders are used for floor beams in buildings. Open web girders are attractive in that they are light and permit the easy installation of services. Fire protection is provided by the floor slab and ceiling or by profile asbestos spray on the steel sections. See Section 2.5.

(e) Columns

Universal columns, compound and built up sections and circular and box sections are used. External columns may be in the facade or placed internally or externally to it. Columns must be provided with fire protection in solid or hollow construction. In other countries external box columns through which water is circulated to give fire protection are used. Column bases are usually made from slabs but fabricated or built-up bases are also used.

(f) Hangers

Rounds, flats, or sections in high-strength steel or steel cables are used for hangers in suspension structures.

(g) Bracings

Angles, channels, universal beam or column sections or structural hollow sections are used as bracing members. Joints are made with black bolts or high strength friction grip bolts. Bracing must be enclosed between fire-resistant walls.

10.5 Loading

10.5.1 Dead load

(a) Roof

The dead load includes the weight of the slab or deck, insulation, finish, e.g.

the screed, felt, or asphalt topping, the steel beams, ceiling and services. Dead loads range from 2–6 kN/m^2.

(b) Floors
The dead load includes the slab, screed, finish, steel beams, ceiling and services. Dead loads range from 4–7 kN/m^2 depending on the type of construction used and the span of the floor slabs.

(c) Walls
The weights of external walls are as follows:

Curtain walls	= 1.0–2.0 kN/m^2
Glazing – single and double respectively	= 0.3–0.6 kN/m^2
Cavity brick walls – 254 mm thick, plastered one side	= 5 kN m^2.

The weights of internal walls depend on the thickness, materials used and type of construction. Some common examples are:

Non-load-bearing walls 100 mm thick plastered both sides:

brick	= 2.4 kN/m^2
breeze block	= 2.2 kN/m^2.

Moveable partitions – allow 1 kN/m^2 of floor area.

(d) Structural members
The size and weight must be estimated where these have not been included in an overall area weight such as the weight of a floor. The weight must also include fire protection if required.

10.5.2 Imposed load
The imposed load on flat roofs where access is provided is 1.5 kN/m^2. The imposed load on floors depends on the occupancy of the building. These are given in Table 1, CP3, Chapter V, Part 1, for various types of buildings. Note that heavier loads must be used on floors where filing and storage space is provided or data processing equipment is installed. This loading may be reduced on multi-storey buildings in accordance with Table 2, CP3. This reduction is allowed because it is unlikely that all floors will be fully loaded at the same time.

10.5.3 Lifts and services

(a) Lifts
The loading on a building caused by a lift is carried at the top of the lift shaft on columns or the reinforced concrete walls in the core. The load consists of the lift drive motor controls and machinery, the lift car, counterweight and ropes, etc, the load lifted, i.e. people or goods and the inertia loads from starting and stopping.

The total loading is expressed as an equivalent dead load which depends on the capacity of the lift. The lift capacity is the net load hoisted expressed as either the number of persons or weight of goods carried. The equivalent dead load includes all the loads listed above, except the structural support floor slab and members.

Lift manufacturers will supply information on loading and sizes required for lift cars and lift compartments and motor rooms for lift machinery of various capacities.

(b) Escalators
The total load from a loaded escalator is given as an equivalent dead load. The load depends on the width rise and span of the escalator. This information is given in manufacturers literature and also clearance dimensions for installation and the location of supporting beams required.

(d) Services
Other services include lighting, heating, ventilating, air conditioning, water supply for human and industrial use and fire fighting and waste disposal. The locations, space requirements and weights for cables, ducts, chutes, pipes, etc, must be ascertained from the services consultants so that these can be accurately taken into account in the design.

Services in floors are included by an appropriate allowance in the floor dead loads. Services in the shaft compartment in the core require supporting at appropriate floor levels. Large items such as water tanks are at roof level and in intermediate levels in tall buildings. The load depends on the capacity required.

10.5.4 Wind loads
The wind loads are taken from CP3, Chapter V, Part 2. The wind load increases with height in accordance with the factor S_2 for ground roughness, building size and height above ground from Table 3, CP3. From the code the dynamic pressure,

$$q = 0.613 V_s^2,$$

where $V_s = S_1 S_2 S_3 V$ = design wind speed,

   S_1 = topography factor = 1.0,

   S_3 = statistical factor   = 1.0,

   V = basic wind speed, which depends on the location.

The variation in the factor S_2, with height for ground roughness 4 and building size Class C, i.e. where either the greatest horizontal or vertical dimension exceeds 50 m, is shown in Fig. 10.9(a).

The wind force = CqA

where C is the pressure coefficient where either of the following values which depend on the building dimensions can be used:

(i) C_{pe} for the walls of rectangular clad buildings from Table 7, CP3, where the values for the windward and leeward faces are added;

(ii) C_f the force coefficient from Table 10, CP3.

A is the area of building surface considered. The wind loads are applied at floor levels as shown in Fig. 10.9(b). The uplift on the roof is usually neglected.

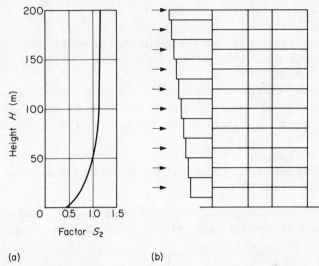

(a)   (b)

Figure 10.9 (a) Variation in wind load and height. (b) Wind loads on a tall building.

10.6 Analysis

10.6.1 Braced and core structures
The steel frames for braced buildings and for core structures are statically determinate and manual methods may be used for the analysis as noted above.

10.6.2 Rigid frame structures
For rigid frame structures the exact manual methods, such as moment distribution are too tedious to apply. Special variations have been developed which can be useful in some cases. The reader should consult Lightfoot [5] and the *Steel Designers Manual* [7] for further information. Computer analysis using standard plane frame or space frame programs can be applied to all cases.

To obtain maximum moments the imposed load must be applied in the appropriate patterns to give maximum hogging and sagging moments and shears in the floor beams and maximum column loads. In general three patterns may need investigating. These are:

(i) imposed load over the whole floor;
(ii) imposed load applied to adjacent spans;
(iii) imposed load applied to alternate spans.

This can lead to the analysis of many load cases.

In addition to the above rigorous methods it is often useful to be able to apply approximate methods of analysis. These are discussed briefly below.

(a) Vertical loads

For vertical loads the whole frame can be analysed by moment distribution. The result is accurate if the frame and loads are symmetrical. Multi-storey frames are usually symmetrical.

Alternatively, sub-frames consisting of the beam at one level and columns above and below that level with ends remote from the beam fixed may be analysed. It may be necessary to consider various patterns for the imposed load.

(b) Horizontal loads

The method of analysis for horizontal loads, termed the portal method, is based on the following assumptions:

(i) the storey shears are divided between the bays in proportion to their widths;
(ii) points of contraflexure are assumed at the mid points of beams and columns.

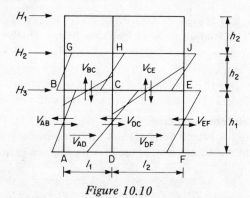

Figure 10.10

The method is explained through reference to Fig. 10.10 which shows a rigid frame loaded with horizontal loads at floor levels. Consider the first storey.

Storey shear $V = H_1 + H_2 + H_3$

Divide this between the bays AD and DF:

Bay AD $V_{AD} = Vl_1/(l_1 + l_2)$

Bay DF $V_{DF} = Vl_2/(l_1 + l_2).$

The bay shears are distributed to the columns:

Column AB $\quad V_{AB} = \frac{1}{2} V_{AD}$

Column DC $\quad V_{DC} = \frac{1}{2} [V_{AD} + V_{DF}]$

Column FE $\quad V_{FE} = \frac{1}{2} V_{DE}$.

The column moments are then calculated:

$$M_{AB} = M_{BA} = \frac{1}{2} V_{AB} h_1$$

$$M_{DC} = M_{CD} = \frac{1}{2} V_{DC} h_1$$

$$M_{FE} = M_{EF} = \frac{1}{2} V_{FE} h_1.$$

The beam moments are calculated:

$$M_{BC} = M_{BA} + M_{BG} = M_{CB}$$

$$M_{EC} = M_{EF} + M_{EJ} = M_{CE}.$$

For equilibrium at joint C:

$$M_{CB} + M_{CE} = M_{CD} + M_{CH}.$$

The shears in the floor beams V_{BC} and V_{CE} are equal in value and only the external columns carry axial load.

10.7 Simple design — multi-storey braced building

10.7.1 Specification

The framing plans and front facade for a five-storey college building are shown in Fig. 10.11(a) and (b). The construction details at columns and floor beams are shown in Fig. 10.12. The roof and floors are cast *in situ* one-way reinforced concrete slabs with screed on top. Suspended ceilings are provided below the floor beams. The external walling consists of curtain wall of cladding panels, breeze block, insulation and lining and double glazed window areas. Breeze block walls are provided around the lift shaft, stairs and toilet blocks. Double leaf walls are required in the braced bays. All other internal walling is in lightweight partitions. The external columns are protected in lightweight casing. Referring to the Table to Regulation E5 of *The Building Regulations*, the building is in purpose group II, the minimum period of fire resistance for structural members above ground is 1 h where the building is less than 28 m in height and the floor area less than 2000 m^2.

The following design work is to be carried out. See Fig. 10.11(a).

(a) Design roof and floor beams, mark numbers 1, 2, 3, 4, 5, 6.
(b) Design columns A1 and A2.
(c) Design the bracing between columns A2 and A3.

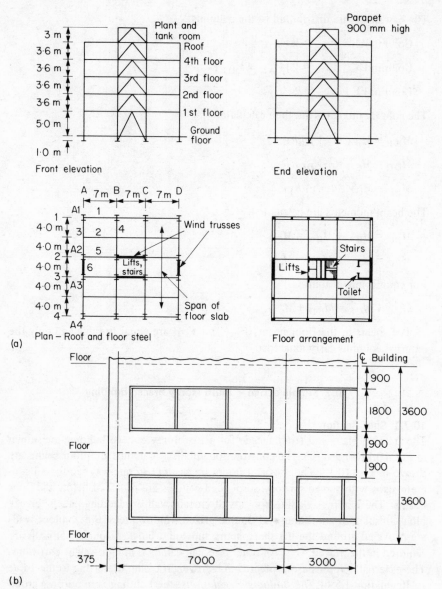

Figure 10.11 (a) Framing plans. (b) Facade – front elevation.

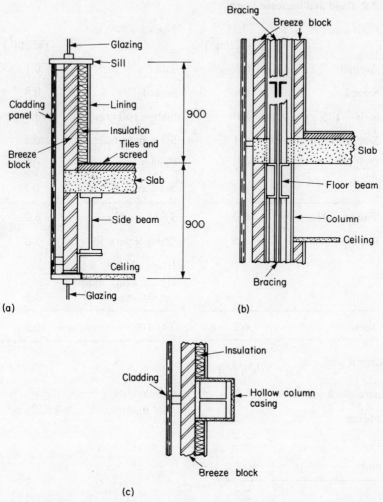

Figure 10.12 (a) Section through external wall. (b) External wall at braced bay. (c) Section through column.

The design for liftshaft and stairwell columns has been omitted for the sake of brevity. For this design the weights of lifts, machinery, tank, stair slabs and shaft walling are required. The material to be used throughout is Grade 43 steel.

10.7.2 Dead and imposed loads

Roof loads	(kN/m²)	*Floor loads*	(kN/m²)
Asphalt	0.5	Tiles	0.1
Screed	0.6	Screed	0.8
Slab — 125 mm	3.0	Slab — 160 mm	3.8
Steel and protection	0.4	Steel and protection	0.5
Ceiling	0.5	Ceiling	0.7
		Services	0.1
Total dead load	5.0	Total dead load	6.0
Imposed load	1.5	Internal partitions	1.0
		Imposed load (computing equipment, etc.)	3.5
Total	6.5	Total	10.5

External wall	(kN/m²)	
Breeze block 100 mm	2.2	Double glazing
Cladding	0.2	and mullions = 0.6 kN/m²
Insulation	0.2	
Lining	0.2	
Total	2.8	

Double skin wall at braced bay (parapet wall same)

	(kN/m²)
Breeze block — 2/100 mm	4.4
Cladding	0.2
Insulation	0.2
Lining	0.2
Total	5.0

Blockwall walls to toilet block

Breeze block 200 m plastered both sides $= 4.6 \text{ kN/m}^2$.

Columns — casing and external walling:

Roof to 2nd floor $= 2.4 \text{ kN/m}$

Second floor to base $= 3.0 \text{ kN/m}$.

Bracing add to appropriate column $= 0.2 \text{ kN/m}$.

The above loading is modified in particular cases where required for the design of the structural members. Some simplifying approximations are made.

10.7.3 Wind loads and wind truss analysis
These are in accordance with CP3, Chapter V, Part 2. The building is located in the suburbs of a city in the North East of England.

Basic wind speed $V = 45$ m/s

Topography factor $S_1 = 1.0$

Ground roughness — category 3

Building size — class B

The wind loads are applied at the floors of the building. The height for determining the S_2 factors, the factors, design wind speeds, dynamic pressures are shown in Fig. 10.13(a) and (b). The derivation of the force coefficient from Table 10, CP3, is given in Fig. 10.13(c) and the calculations for the forces at roof and floor levels are given in Fig. 10.13(d). The wind truss consisting of bracing and columns is analysed by joint resolution. The forces in the members are shown in Fig. 10.14.

10.7.4 Beam design
The design for the roof and floor beams is set out in tabulated form below. Figures showing the loading on the beams are included in the calculations. Some notes on the design procedure are:

(a) all beams are considered to be fully restrained against lateral buckling by friction between the top flange and slab;
(b) the deflection of all beams subjected to their total load will be limited to span/360. Then the moment of inertia required for uniformly distributed loading is

$$I = 2.34 \, WL^2 \text{ cm}^4,$$

where W is the total uniformly distributed load (kN), and L is the span (m).
For a beam with a central load W, $I = 3.75 \, WL^2 \text{ cm}^4$;
(c) the shear stress is not checked as this is very small.

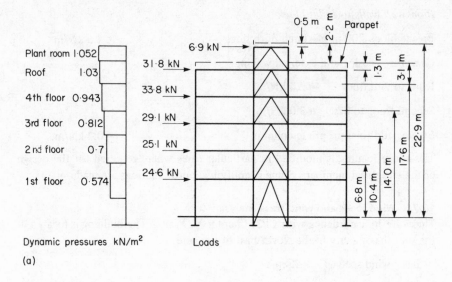

Dynamic pressures kN/m²

(a)

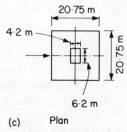

(c)   Plan

Figure 10.13   (a) Wind pressures and building loads. (b) Calculation of dynamic pressures. (c) Force coefficients. (d) Wind loads at roof and floor levels.

10.7.5 Column design

(a) Corner column Al

The columns will be fabricated in two lengths and spliced at second floor level. Design calculations are given for the critical sections in each length only. The column, roof and floor beam reactions, the trial sections and properties are shown in Fig. 10.15.

Roof to second floor   203 x 203 UC 60

Below third floor level – Imposed load reduction	= 20%
Dead load = 64.4 + 67.5 + 2(83 + 85.8) + (2 x 8.6)	= 486.7 kN
Imposed load = 0.8 [(2 x 10.15) + (4 x 24.5)]	= 95.2 kN
Total axial load	= 581.9 kN.

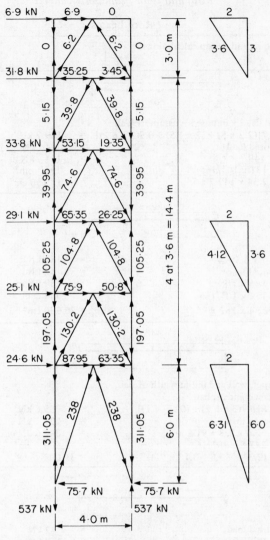

Figure 10.14

Moments $M_{XX} = 0.5(85.8 + 0.8 \times 24.5)0.205$ = 10.8 kN m

$M_{YY} = 0.5(83 + 0.8 \times 24 \times 5)0.105$ = 5.4 kN m.

Slenderness ratio $l/r_{YY} = 0.85 \times 3600/51.9$ = 59.

Allowable stresses: p_c = 126 N/mm^2

p_{bc} = 165 N/mm^2

Roof beams

Mark	Loading and design calculations	Section and properties
1	Roof dead + imposed + parapet + wall $W = 7[(2.2 \times 5) + (2 \times 1.5) + 0.9(3 + 5.2)]$ $=$ 149.7 kN Imposed load $=$ 21 kN $M = 149.7 \times 7/8$ $=$ 131 kN m $Z = 131 \times 10^3/165$ $=$ 794 cm³ $I = 2.34 \times 149.7 \times 7^2$ $=$ 17 150 cm⁴	406 × 178 UB 54 $Z = 922.8$ cm³ $I = 18\,576$ cm⁴
2 5	$W = 7 \times 6.5 \times 4$ $=$ 182 kN Imposed load $=$ 42 kN $M = 182 \times 7/8$ $=$ 159 kN m $Z = 159 \times 10^3/165$ $=$ 964 cm³ $I = 2.34 \times 182 \times 7^2$ $=$ 20 900 cm⁴	457 × 152 UB 60 $Z = 1120$ cm³ $I = 25\,464$ cm⁴
3	Parapet, wall and 0.2 m width of slab, screed and asphalt $W = 8[0.9(3 + 5.2) + (0.2 \times 4.1)]$ $=$ 65.6 kN Imposed load $=$ 21 kN $M = 65.6 \times 8/8 + 91 \times 8/4$ $=$ 247.6 kN m $Z = 247.6 \times 10^3/165$ $=$ 1 500 cm³ $I = (2.34 \times 65.6 + 3.75 \times 91)8^2$ $=$ 31 600 cm⁴	533 × 210 UB 82 $Z = 1\,793$ cm³ $I = 47\,363$ cm⁴
4	Imposed load $=$ 42 kN $M = 182 \times 8/4$ $=$ 362 kN $Z = 362 \times 10^3/165$ $=$ 2 195 cm³ $I = 3.75 \times 182 \times 8^2$ $=$ 43 500 cm⁴	610 × 229 UB 101 $Z = 2\,509$ cm³ $I = 75\,549$ cm⁴
6	Treat as a continuous beam Double skin wall + parapet + slab + beam $W = 4[(1.8 \times 5) + (0.2 \times 4) + 0.5]$ $=$ 41.2 kN See design for floor beam Mark 6	2/178 × 76] [20.84

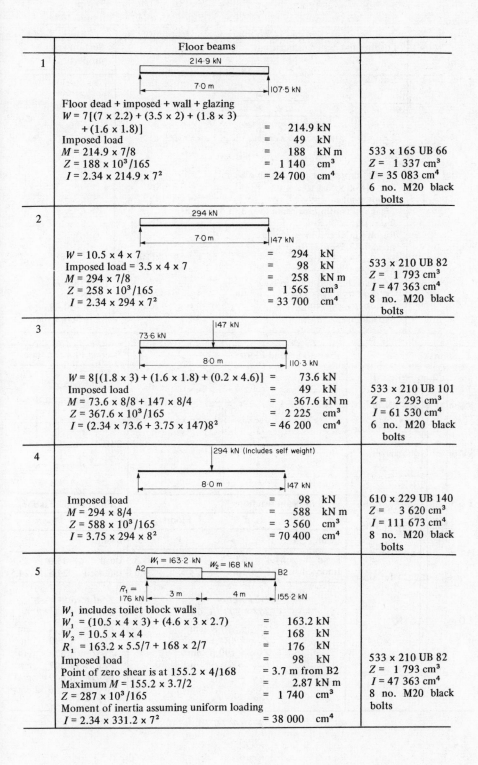

	Floor beams	
1	**214·9 kN** 7·0 m — 107·5 kN Floor dead + imposed + wall + glazing $W = 7[(7 \times 2.2) + (3.5 \times 2) + (1.8 \times 3)$ $\quad + (1.6 \times 1.8)]$ = 214.9 kN Imposed load = 49 kN $M = 214.9 \times 7/8$ = 188 kN m $Z = 188 \times 10^3/165$ = 1 140 cm³ $I = 2.34 \times 214.9 \times 7^2$ = 24 700 cm⁴	533 × 165 UB 66 $Z = 1\ 337$ cm³ $I = 35\ 083$ cm⁴ 6 no. M20 black bolts
2	**294 kN** 7·0 m — 147 kN $W = 10.5 \times 4 \times 7$ = 294 kN Imposed load = $3.5 \times 4 \times 7$ = 98 kN $M = 294 \times 7/8$ = 258 kN m $Z = 258 \times 10^3/165$ = 1 565 cm³ $I = 2.34 \times 294 \times 7^2$ = 33 700 cm⁴	533 × 210 UB 82 $Z = 1\ 793$ cm³ $I = 47\ 363$ cm⁴ 8 no. M20 black bolts
3	73·6 kN ↓147 kN 8·0 m — 110·3 kN $W = 8[(1.8 \times 3) + (1.6 \times 1.8) + (0.2 \times 4.6)]$ = 73.6 kN Imposed load = 49 kN $M = 73.6 \times 8/8 + 147 \times 8/4$ = 367.6 kN m $Z = 367.6 \times 10^3/165$ = 2 225 cm³ $I = (2.34 \times 73.6 + 3.75 \times 147)8^2$ = 46 200 cm⁴	533 × 210 UB 101 $Z = 2\ 293$ cm³ $I = 61\ 530$ cm⁴ 6 no. M20 black bolts
4	↓294 kN (Includes self weight) 8·0 m — 147 kN Imposed load = 98 kN $M = 294 \times 8/4$ = 588 kN m $Z = 588 \times 10^3/165$ = 3 560 cm³ $I = 3.75 \times 294 \times 8^2$ = 70 400 cm⁴	610 × 229 UB 140 $Z = 3\ 620$ cm³ $I = 111\ 673$ cm⁴ 8 no. M20 black bolts
5	$W_1 = 163\cdot2$ kN $W_2 = 168$ kN A2 _____ B2 $R_1 = 176$ kN — 3 m — 4 m — 155·2 kN W_1 includes toilet block walls $W_1 = (10.5 \times 4 \times 3) + (4.6 \times 3 \times 2.7)$ = 163.2 kN $W_2 = 10.5 \times 4 \times 4$ = 168 kN $R_1 = 163.2 \times 5.5/7 + 168 \times 2/7$ = 176 kN Imposed load = 98 kN Point of zero shear is at $155.2 \times 4/168$ = 3.7 m from B2 Maximum $M = 155.2 \times 3.7/2$ = 2.87 kN m $Z = 287 \times 10^3/165$ = 1 740 cm³ Moment of inertia assuming uniform loading $I = 2.34 \times 331.2 \times 7^2$ = 38 000 cm⁴	533 × 210 UB 82 $Z = 1\ 793$ cm³ $I = 47\ 363$ cm⁴ 8 no. M20 black bolts

Mark	Loading and design calculations	Section and properties
6	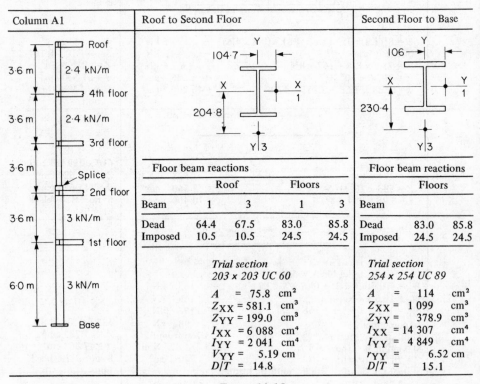 $W = 77.6$ kN A2　87·95　X　63·35　A3 14·53 kN↑　　　　↑14·53 kN $238 - 25.7 = 212.3$ kN↗　↖$238 + 25.7 = 263.7$ kN $R_x = 48.5$ kN Double skin wall + part of slab + beam $W = 4[(3.6 \times 5) + (0.2 \times 4.6) + 0.5]$ = 77.6 kN Imposed load = 0 Treat as a continuous beam load 38.8 kN/span $R_X = 1.25 \times 38.8$ = 48.5 kN Resolve in direction of bracing Load = $48.5 \times 6.31/(6.0 \times 2)$ = 25.7 kN $M = 0.125 \times 38.8 \times 2$ = 9.7 kN m Axial load = 87.95 kN Try two channels back to back $f_c = 87.95 \times 10/53.08$ = 16.6 N/mm² $f_{bc} = 9.7 \times 10^3/300.8$ = 32.3 N/mm² Satisfactory.	2/178 × 76] [20.84 $A = 53.08$ cm² $Z = 300.8$ cm³

Column A1	Roof to Second Floor	Second Floor to Base
3·6 m　Roof　2·4 kN/m 　　4th floor 3·6 m　2·4 kN/m 　　3rd floor 3·6 m 　Splice 　2nd floor 3·6 m　3 kN/m 　1st floor 6·0 m　3 kN/m 　Base	104·7→ ← X　　　X 1 204·8 Y 3	106→ ← X　　　Y 1 230·4 Y 3

Floor beam reactions					Floor beam reactions	
	Roof		Floors		Floors	
Beam	1	3	1	3	Beam	
Dead	64.4	67.5	83.0	85.8	Dead	83.0 85.8
Imposed	10.5	10.5	24.5	24.5	Imposed	24.5 24.5

Trial section
203 × 203 UC 60

$A = 75.8$ cm²
$Z_{XX} = 581.1$ cm³
$Z_{YY} = 199.0$ cm³
$I_{XX} = 6\,088$ cm⁴
$I_{YY} = 2\,041$ cm⁴
$V_{YY} = 5.19$ cm
$D/T = 14.8$

Trial section
254 × 254 UC 89

$A = 114$ cm²
$Z_{XX} = 1\,099$ cm³
$Z_{YY} = 378.9$ cm³
$I_{XX} = 14\,307$ cm⁴
$I_{YY} = 4\,849$ cm⁴
$r_{YY} = 6.52$ cm
$D/T = 15.1$

Figure 10.15

Actual stresses: $f_c = 581.9 \times 10/76.8$ $= 76.8 \text{ N/mm}^2$

XX axis $f_{bc} = 10.8 \times 10^3/581.1$ $= 18.6 \text{ N/mm}^2$

YY axis $f_{bc} = 5.4 \times 10^3/199$ $= 27.2 \text{ N/mm}^2.$

Combined: $\dfrac{f_c}{p_c} + \dfrac{f_{bc}}{p_{bc}} = \dfrac{76.8}{126} + \dfrac{18.6 + 27.2}{165}$ $= 0.89.$

The column section is satisfactory.

Second floor to base   254 x 254 UC 89

Below first floor – Imposed load reduction $= 40\%$

Dead load $= 64.4 + 67.5 + 4(83 + 85.8) + (3 \times 8.6) + 10.8$ $= 843.7 \text{ kN}$

Imposed load $= 0.6 \ [(2 \times 10.5) + (8 \times 24.5)]$ $= 130.2 \text{ kN}$

Total axial load $= 973.9 \text{ kN}.$

Distribution factors:

Column base to 1st floor $= \dfrac{1}{6} \ \left(\dfrac{1}{3.6} + \dfrac{1}{6} \right)$ $= 0.382.$

Moments: $M_{XX} = 0.382(85.8 + 0.6 \times 24.5)0.23$ $= 9.65 \text{ kN m}$

$M_{YY} = 0.382(83 + 0.6 \times 24.5)0.106$ $= 3.97 \text{ kN m}.$

Slenderness ratio $l/r_{YY} = 0.85 \times 6000/65.2$ $= 78.3.$

Allowable stresses: p_c $= 105.7 \text{ N/mm}^2$

p_{bc} $= 165 \ \text{ N/mm}^2.$

Actual stresses: $f_c = 973.9 \times 10/114$ $= 85.3 \text{ N/mm}^2$

XX axis $f_{bc} = 9.65 \times 10^3/1099$ $= 8.8 \text{ N/mm}^2$

YY axis $f_{bc} = 3.97 \times 10^3/378.9$ $= 10.5 \text{ N/mm}^2$

Combined: $\dfrac{f_c}{p_c} + \dfrac{f_{bc}}{p_{bc}} = \dfrac{85.3}{105.7} + \dfrac{8.8 + 10.5}{165}$ $= 0.925.$

The column may be checked at other points. The section selected is satisfactory.

(b) Column A2

The column, roof and floor beam reactions, wind loads and trial sections and properties are shown in Fig. 10.16. The column is spliced at second floor level.

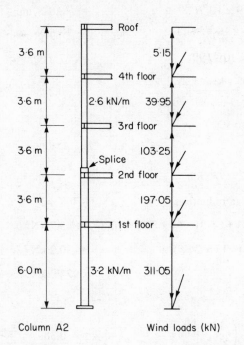

Column A2   Wind loads (kN)

Figure 10.16

Roof to second floor – 254 x 254 UC 73

Below third floor level – Imposed load reduction	= 20%
Dead load = 67.8 + 70.0 + 21.6 + 2(85.8 + 127 + 38.8) + (7 x 2.6)	= 678.8 kN
Imposed load = 0.8 [10.5 + 21.0 + 2(24.5 + 49.0)]	= 142.8 kN
Total	= 821.6.

Wind load = 105.25 kN, i.e. 12.8% of total load. Neglect.

Moments: $M_{XX} = 0.5[85.8 - 14.6) + (0.8 \times 24.5)]\,0.227$ = 10.3 kN m

$\qquad\qquad M_{YY} = 0.5\,[127 + (0.8 \times 49)]\,0.104$ = 8.7 kN m.

Slenderness ratio $l/r_{YY} = 0.7 \times 3600/64.6$ = 39.

Allowable stress: p_c = 139 N/mm^2

$\qquad\qquad\qquad p_{bc}$ = 165 N/mm^2.

Actual stresses: $f_c = 821.6 \times 10/92.9$ $= 87.6 \text{ N/mm}^2$

XX axis $f_{bc} = 10.3 \times 10^3/894.5$ $= 11.5 \text{ N/mm}^2$

YY axis $f_{bc} = 8.7 \times 10^3/305$ $= 28.6 \text{ N/mm}^2.$

Combined: $\dfrac{f_c}{p_c} + \dfrac{f_{bc}}{p_{bc}} = \dfrac{87.6}{139} + \dfrac{11.5 + 28.6}{165}$ $= 0.873.$

The column section is satisfactory.

Second floor to base – 254 x 254 UC 132

Below first floor – Imposed load reduction $= 40\%$

Dead load $= 67.8 + 70 + 21.61 + 4(85.8 + 127 + 38.8) +$
    $(3 \times 3.6 \times 2.6) + (3.6 \times 3.2)$ $= 1205.4 \text{ kN}$

Imposed load $= 0.6[10.5 + 21.0 + 4(24.5 + 49)]$ $= 195.3 \text{ kN}$

Total load $= 1400.7 \text{ kN.}$

Wind load $= 311.05$ kN, i.e. 22.2% of total load. Neglect.

Distribution factors:

Column to base to 1st floor $= 0.382$

Moments: $M_{XX} = 0.382[85.8 - 14.6 + 0.6 \times 24.5]0.238$ $= 7.81 \text{ kN m}$

$M_{YY} = 0.382[127 + 0.6 \times 49]0.106$ $= 6.35 \text{ kN m.}$

Slenderness ratio $l/r_{YY} = 0.85 \times 6000/66.6$ $= 76.6.$

Allowable stresses: p_c $= 107.4 \text{ N/mm}^2$

p_{bc} $= 165 \text{ N/mm}^2.$

Actual stresses: $f_c = 1400.7 \times 10/167.7$ $= 83.5 \text{ N/mm}^2$

XX axis $f_{bc} = 7.81 \times 10^3/1622$ $= 4.82 \text{ N/mm}^2$

YY axis $f_{bc} = 6.35 \times 10^3/570.4$ $= 11.1 \text{ N/mm}^2.$

Combined: $\dfrac{f_c}{p_c} + \dfrac{f_{bc}}{p_{bc}} = \dfrac{83.5}{107.4} + \dfrac{4.82 + 11.1}{165}$ $= 0.874.$

(c) Column base plates

Column A1

Load $= 973.9 + 6 \times 3$ $= 991.9 \text{ kN.}$

Permissible bearing stress on the concrete $= 4 \text{ N/mm}^2.$

Area of base = $991.9 \times 10^3/4$ $= 2.48 \times 10^5$ mm^2.

Provide a slab 500 mm x 500 mm.

Bearing pressure $= 3.97$ N/mm^2.

The base is shown in Fig. 10.17(a).

The thickness of base plate

$$t = \left[\frac{3 \times 3.97}{185} \left(122.1^2 - \frac{119.8^2}{4} \right) \right]^{1/2}$$ $= 26.99$, say 30 mm.

Base plate 500 m x 500 m x 30 mm thick with four no. 24 mm diameter H.D. bolts.

Column A2
Load = $1400.7 + 6 \times 3$ $= 1418.7$ kN.

Area of base = $1418.7 \times 10^3/4$ $= 3.55 \times 10^5$ mm^2.

Provide a slab 600 mm x 600 mm.

Bearing pressure $= 3.94$ N/mm^2.

The base is shown in Fig. 10.17(b).

The thickness of base plate:

$$t = \left[\frac{3 \times 3.94}{185} \left(169.5^2 - \frac{161.8^2}{4} \right) \right]^{1/2}$$ $= 37.7$, say 40 mm.

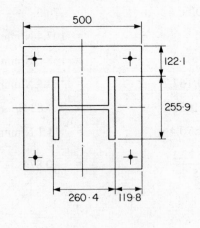

Base plate 500 x 500 x 30 thick   Base plate 600 x 600 x 40 thick
HD bolts four no. M24   HD bolts four no. M24
Column A1   Column A2

Figure 10.17

Base plate 600 mm x 600 mm x 40 mm thick with four no. 24 mm diameter H.D. bolts.

10.7.6 Bracing design
The bracing size will be changed at second floor level.

(a) Roof to second floor

Load = wind load + reaction from beam Mark 6

$\qquad$ = 104.8 + 51.5 $\qquad$ = 156.3 kN.

Try two no. 100 x 65 x 8 ⌐L $\qquad$ A = 25.3 cm^2; $\qquad$ r = 2.75 cm

l/r = 4120 x 0.85/27.5 $\qquad$ = 128

p_c = 53 + 25% $\qquad$ = 66.2 N/mm^2

f_c = 156.3 x 10/25.3 $\qquad$ = 62.0 N/mm^2.

Gusset between legs, one no. 22 mm diameter hole.

Safe load in tension $\qquad$ = 338 kN.

Provide two no. 100 x 65 x 8 ⌐L.

M20, HSFG bolts, single shear value including wind $\qquad$ = 54 kN.

The bolts are in double shear. Provide two bolts each end.

(b) Second floor to base

Load = wind load + reaction from beam Mark 6

$\qquad$ = 238 + 51.5 $\qquad$ = 289.5 kN.

Try two no. 150 x 90 x 12 ⌐L. $\qquad$ A = 55 cm^2, $\qquad$ r = 3.69 cm

l/r = 6310 x 0.85/36.9 $\qquad$ = 145

p_c = 43 + 25% $\qquad$ = 53.7 N/mm^2

f_c = 289.5 x 10/55 $\qquad$ = 52.6 N/mm^2.

Provide two no. 150 x 90 x 12 ⌐L.

M20, HSFG bolts. Provide three bolts each end.

10.7.7 Steelwork details
Framing members and typical steelwork details are shown in Fig. 10.18.

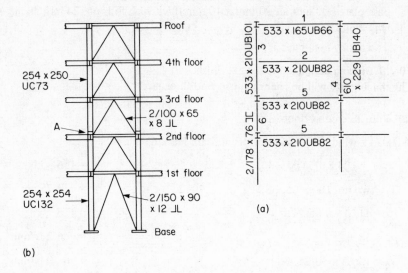

(a)

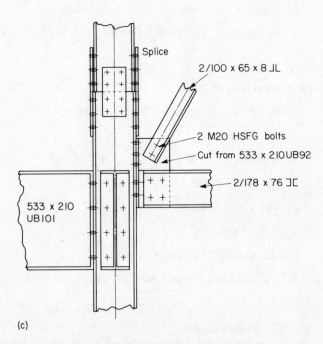

(b)

(c)

Figure 10.18 (a) Part floor plan. (b) Elevation at braced bay. (c) Detail at A.

10.8 Coursework exercises

1. The framing plans for a four storey office building are shown in Fig. 10.19. The design loading is as follows:

(i) Roof: Dead load — screed, finish, slab, steel,
   ceiling, services $= 5$ kN/m^2

    Imposed load $= 1.5\ kN/m^2$.

(ii) Floors: Dead load — screed, finish slab, steel,
   ceiling, services, partitions $= 6.5\ kN/m^2$

    Imposed load $= 2.5\ kN/m^2$.

(iii) Walls: Curtain walling and glazing, steel

    Linear load at each floor $= 7.5\ kN/m$

    Linear load at roof level $= 3.0\ kN/m$.

(iv) Columns: Lightweight casing, allow $= 1.8\ kN/m$.

    Carry out the following design work.

(a) Design the roof and floor steel and the plate girder at first floor level.
(b) Design column, A, B and C.

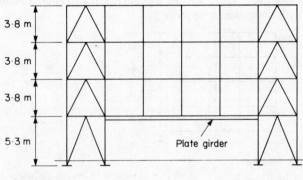

Front elevation

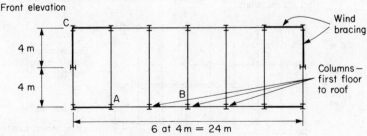

Plan — Roof and floors

Figure 10.19

(c) Design the wind bracing.

(d) Show the main structural details on sketches.

2. The plans and elevation for a four storey building are shown in Fig. 10.20. The loading is as follows:

Roof: Dead load — slab, finish, steel, ceiling, services $= 4.5 \text{ kN/m}^2$

Imposed load $= 1.5 \text{ kN/m}^2$.

Floor: Dead load — slab, finish, tiles, steel, ceiling,
    services, partitions $= 6.5 \text{ kN/m}^2$

Imposed load $= 3.5 \text{ kN/m}^2$.

Corridor walls: breeze block plastered both sides $= 6.0 \text{ kN/m}$.

External cladding: glazing, breeze block, cladding,
    insulation, lining, steel $= 6.0 \text{ kN/m}$.

Wind loads CP3, Chapter V, Part 2. The location of the building is near the city centre.

    Design an internal transverse frame as a rigid frame to carry vertical loads and wind loads. Analyse the frame using a computer program.

3. Design the same frame using the simple design method to carry the vertical loads only. Design the bracing for the end bays. Make an approximate weight comparison with the design in Exercise 2.

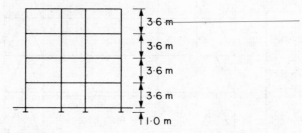

Section

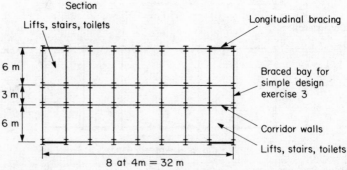

Plan

Figure 10.20

References and further reading

[1] British Steel Corporation (1972). *Building with Steel. Theme: Multi-storey Construction*. November 1972.
[2] BS449, Part 2 (1969). *The Use of Structural Steel in Building*. British Standards Institution, London.
[3] Hart, F., Henn, W. and Sontag, H. (1978). *Multi-storey buildings in steel*. Granada Publishing, London.
[4] Khan, F. R. and Amin, N. R. (1973). Analysis and Design of Framed Tube Structures for Tall Concrete Buildings. *The Structural Engineer* 51 (3).
[5] Lightfoot, E. (1961). *Moment Distribution*. E. & F. N. Spon, London.
[6] Schueller, W. (1977). *High-rise Building Structures*. John Wiley, New York.
[7] *Steel Designers Manual* (1972). Constrado, Crosby Lockwood, London.
[8] *Steel Frame Buildings* (1978). Conder and British Steel Corporation, Winchester.

Projects

A list of suggestions for projects is given below. The projects are meant to be open ended. Constraints should be set by the student in consultation with his supervisor. In general, the projects should consist of:

(a) literature surveys;
(b) drawings showing various proposals;
(c) analysis and design, programming or model construction and testing;
(d) summary of results;
(e) presentation of conclusions.

1. Grandstand buildings.
   Visit football grounds and classify various structural systems which are used. Make comparative design studies for two proposals.
2. Church structure.
   Formulate proposals for using structural steelwork to frame a church building. Make preliminary designs for various schemes.
3. Computer design of portals.
   Set out in detail a scheme for the analysis and design by computer of single bay portals for both the elastic and plastic methods. Write and validate programs to cover as much of the process as can be done with computing facilities available.
4. Model domes.
   Make models using perspex, metal or timber of the following dome types:

   (a) rigid Schwedler dome;
   (b) three-way grid dome;
   (c) geodesic dome based on a triangulated dodecahedron. Here each pentagonal face is to be divided into five triangles where the common apex lies on the supersphere. Use about one-half of the supersphere for the dome.

   Apply various loads to the domes, measure deflections at the joints and strains where possible with electric strain gauges. Check results against calculated values from a computer analysis.
5. Visit an outstanding steel framed structure such as a leisure centre or multi-storey building that has been recently completed in your area. Read any

technical papers that have been published on the structure and contact the owners, architects or engineers involved on the project. It may be possible to obtain further particulars. Make idealized sketches of the structure. Discuss the methods used for analysis and design. Carry out some preliminary calculations to prove sizes of main members. Prepare an alternative scheme for the same structure. Finally, discuss any particular problems with regard to foundations, fabrication and erection.

Index